Lecture Notes in Educational Technology

Series Editors

Ronghuai Huang, Smart Learning Institute, Beijing Normal University, Beijing, China

Kinshuk, College of Information, University of North Texas, Denton, TX, USA

Mohamed Jemni, University of Tunis, Tunis, Tunisia

Nian-Shing Chen, National Yunlin University of Science and Technology, Douliu, Taiwan

J. Michael Spector, University of North Texas, Denton, TX, USA

The series Lecture Notes in Educational Technology (LNET), has established itself as a medium for the publication of new developments in the research and practice of educational policy, pedagogy, learning science, learning environment, learning resources etc. in information and knowledge age, – quickly, informally, and at a high level.

Abstracted/Indexed in:

Scopus, Web of Science Book Citation Index

More information about this series at https://link.springer.com/bookseries/11777

Francisco José García-Peñalvo ·
Alicia García-Holgado · Angeles Dominguez ·
Jimena Pascual
Editors

Women in STEM in Higher Education

Good Practices of Attraction, Access
and Retainment in Higher Education

 Springer

Editors
Francisco José García-Peñalvo 🆔
GRIAL Research Group
University of Salamanca
Salamanca, Spain

Alicia García-Holgado 🆔
GRIAL Research Group
University of Salamanca
Salamanca, Spain

Angeles Dominguez 🆔
Tecnologico de Monterrey
Institute for the Future of Education
Monterrey, Nuevo León, Mexico

Jimena Pascual 🆔
Pontificia Universidad Católica de
Valparaíso
Valparaíso, Chile

ISSN 2196-4963   ISSN 2196-4971   (electronic)
Lecture Notes in Educational Technology
ISBN 978-981-19-1551-2   ISBN 978-981-19-1552-9   (eBook)
https://doi.org/10.1007/978-981-19-1552-9

This Springer imprint is published by the registered company Springer Nature Singapore Pte Ltd.
The registered company address is: 152 Beach Road, #21-01/04 Gateway East, Singapore 189721,
Singapore

Foreword

W-STEM Project

Gender equality must be central to how all our public organisations operate. Gender equality metrics must be transparent. Gender equality demands nothing less. Gender equality benefits everyone—men, women, students and society.

As the first minister in Ireland in charge of Higher Education, I wanted to implement change that outlived my tenure and that was a legacy for our young daughters, our granddaughters and for all the female students in the future who will aspire to a higher education in fair and equitable institutions.

Gender inequality exists in our third-level institutions not because of a lack of talent or ambition, but because of systemic barriers and a culture that means talent alone is not always enough to guarantee success. Gender inequality is discrimination, pure and simple, and I was committed to ensuring that this wrongful practice of discriminating against the appointment of senior female academics stopped within our publicly funded Higher Education Institutions.

I acknowledge the impressive body of work accomplished by Máire Geoghegan-Quinn and her team in the National Review of Gender Equality in Higher Education (2016). Although that report included sixty-six recommendations for action, there was unfortunately no real driver for change within the institutions. I used that report as a springboard to devise a clear and achievable road map and action plan on how we would provide transformative change within our Third Level Institutions. In November 2017, I established a Gender Equality Taskforce which gave a much greater emphasis to the work of achieving gender equality in our higher education institutions. I also want to acknowledge the work done by Department of Education and Skills officials and the Attorney General that allowed me to vehemently argue the case with my government colleagues and get cabinet approval.

A ground-breaking initiative I introduced which I am incredibly proud of is the Senior Academic Leadership Initiative (SALI). It was a blunt message to the Irish Higher Education institutions: Get you act together. It rocked boats especially the luxury liners of the male hierarchies.

The SALI female only initiative ensured that forty-five qualified, competent female academic staff could compete for professorial tenure without conscious and unconscious bias, within a three-year time frame.This breaks down gender role stereotyping, and students in disciplines traditionally underrepresented by women can now see role models that they can aspire to.

The achievement of gender equality requires a genuine commitment and I believe my department officials and I delivered on this.

I hope my work in this area resonates for years. For me rocked boats are frankly irrelevant especially when it came to the righting of an obvious and longstanding wrong.

Mary Mitchell O'Connor
Minister of State for Higher Education
(2017–2020)
Dublin, Ireland

Preface

Equity, access and democratisation of Higher Education are targets related to the sustainable development goals (Inter-Agency and Expert Group on SDG Indicators, 2017). Although the fourth goal focuses on gender equality, ensuring inclusive and equitable quality education and promoting lifelong learning opportunities for all, gender equality is a cross-cutting objective present in most of the SDGs (Esquivel & Sweetman, 2016; Pradhan, Costa, Rybski, Lucht, & Kropp, 2017). The gender gap is a challenge in all sectors of society, but is especially visible in STEM (Science, Technology, Engineering, and Mathematics), both in an educational and professional context (Botella, Rueda, López-Iñesta, & Marzal, 2019; Gomez, Abadía Alvarado, & Bernal Nisperuza, 2020; Quirós et al., 2018; UNESCO. Director-General 2009–2017, 2017; Wang & Degol, 2017; World Economic Forum, 2016). According to the UNESCO Institute for Statistics, the mean percentage of female students in tertiary education enrolled in engineering, manufacturing and construction programs is between 6% and 7% between 2015 and 2018; in contrast, the percentage of male students choosing these careers is around 20–21%.

The result of the underrepresentation group is that many important questions and different points of view are not discussed, losing many important questions and ideas, made by distinct and rich groups that compose the society, that can benefit all (García-Holgado, González-González, & Peixoto, 2020). Moreover, this is also an economic issue; according to the European Union, increasing the number of women in digital jobs will increase the Gross Domestic Product in Europe (Seo, Huang, & Han, 2017).

Organizations across the world are working on reducing the gender gap in STEM, but the situation depends on many factors not only related to cultural and socio-economic context but also factors such as self-perception, self-efficacy or previous educational experiences (Cadaret, Hartung, Subich, & Weigold, 2017; Leaper, & Starr, 2019; Lent, Brown, & Hackett, 1994; Lent, Brown, & Hackett, 2000; Moss-Racusin, Dovidio, Brescoll, Graham, & Handelsman, 2012; Malik, & Al-Emran, 2018; Salami, 2007; Salas-Morera, Cejas Molina, Olivares Olmedilla, García-Hernández, & Palomo-Romero, 2019; Seo, Huang, & Han, 2017). Furthermore, there are many steps across the development of the career where the number of women decreases in these areas: when they enter the university, when they join the labour market and when they achieve high professional positions (Seo, Huang, & Han, 2017; Amon, 2017).

Closing this gender gap requires a holistic approach that involves not only the educational sector (primary, secondary and tertiary education) but also other spheres. It is necessary to face this problem working on the social norms and stereotypes that influence in the perception of how a STEM person should be, the professional progression, the entrepreneurship and innovation sector, the gender balance and gender perspective in research, and the formulation of policies. In this context, higher education institutions have a crucial role in the implementation of measures to reduce the gender gap in STEM. They have an impact on other educational levels through the training of future teachers, and also have an impact in the labour market. Furthermore, there is a need for modernising policies, governance, and strengthening relations between higher education systems and the economic and social environment, including other educational levels, to work in closing the gender gap in STEM.

This book addresses the challenges related to women in STEM in higher education, gathering research, studies, good practices, experiences and so on about the engagement, accessing and retention of women in the STEM disciplines, also the strategies of the universities and policymakers to reduce the existing gender gap in these areas. The chapters provide an overview of the approaches implemented in different regions worldwide and provide numerous examples that can be transferred to other higher education institutions.

Salamanca, Spain   Francisco José García-Peñalvo
Salamanca, Spain   Alicia García-Holgado
Monterrey, Mexico   Angeles Dominguez
Valparaíso, Chile   Jimena Pascual

References

Amon, M. J. (2017). Looking through the glass ceiling: A qualitative study of STEM women's career narratives. *Frontiers in psychology, 8.* doi:https://doi.org/10.3389/fpsyg.2017.00236.

Botella, C., Rueda, S., López-Iñesta, E., Marzal, P. (2019). Gender diversity in STEM disciplines: A multiple factor problem. *Entropy, 21*(1), 30. doi:https://doi.org/10.3390/e21010030.

Cadaret, M. C., Hartung, P. J., Subich, L. M., Weigold, I. K. (2017). Stereotype threat as a barrier to women entering engineering careers. *Journal of Vocational Behavior, 99*, 40–51. doi:https://doi.org/10.1016/j.jvb.2016.12.002.

Esquivel, V., Sweetman, C. (2016). Gender and the sustainable development goals. *Gender & Development, 24*(1), 1–8. doi:https://doi.org/10.1080/13552074.2016.1153318.

García-Holgado. A., González-González, C. S., Peixoto, A. (2020). A comparative study on the support in engineering courses: a case study in Brazil and Spain. *IEEE Access, 8*, 125179–125190. doi:https://doi.org/10.1109/ACCESS.2020.3007711.

Gomez Soler, S. C., Abadía Alvarado, L. K., Bernal Nisperuza, G. L. (2020). Women in STEM: does college boost their performance? *Higher Education, 79*(5), 849–866. doi:https://doi.org/10.1007/s10734-019-00441-0.

Inter-Agency and Expert Group on SDG Indicators. (2017). *Global indicator framework for the sustainable development goals and targets of the 2030 agenda for sustainable development.* United Nations

Leaper, C., Starr, C. R. (2019). Helping and hindering undergraduate women's STEM motivation: Experiences with STEM encouragement, STEM-related gender bias, and sexual harassment. *Psychology of Women Quarterly, 43*(2), 165-183. doi:https://doi.org/10.1177/036168431880 6302.

Lent, R. W., Brown, S. D., Hackett, G. (1994). Toward a unifying social cognitive theory of career and academic interest, choice, and performance. *Journal of Vocational Behavior, 45*(1), 79–122. doi:https://doi.org/10.1006/jvbe.1994.1027.

Lent, R. W., Brown, S. D., Hackett, G. (2000). Contextual supports and barriers to career choice: A social cognitive analysis. *Journal of Counseling Psychology, 47*(1), 36–49. doi:https://doi.org/10.1037/0022-0167.47.1.36.

Malik, S. I., Al-Emran, M. (2018). Social factors influence on career choices for female. *Computer Science Students, 13*(05), 15. doi:https://doi.org/10.3991/ijet.v13i05.8231.

Moss-Racusin, C. A., Dovidio, J. F., Brescoll, V. L., Graham, M. J., Handelsman, J. (2012). Science faculty's subtle gender biases favor male students. *Proceedings of the National Academy of Sciences, 109*(41), 16474–16479. doi:https://doi.org/10.1073/pnas.1211286109.

Pradhan, P., Costa, L., Rybski, D., Lucht, W., Kropp, J. P. (2017). A systematic study of sustainable development goal (SDG) interactions. *Earth's Future, 5*(11), 1169–1179. doi:https://doi.org/10.1002/2017EF000632.

Quirós, C. T., Morales, E. G., Pastor, R. R., Carmona, A. F., Ibáñez, M. S., Herrera, U. M. (2018). *Women in the digital age.* Luxembourg: Publications Office of the European Union. doi:https://doi.org/10.2759/526938.

Salami, S. O. (2007). Influence of culture, family and individual differences on choice of gender-dominated occupations among female students in tertiary institutions. *Women in Management Review, 22*(8), 650–665. doi:https://doi.org/10.1108/09649420710836326.

Salas-Morera, L., Cejas Molina, A., Olivares Olmedilla, J. L., García-Hernández, L., Palomo-Romero, J. M. (2019). Factors affecting engineering students dropout: A case study. *International Journal of Engineering Education, 35*(1), 156–167

Seo, G., Huang, W., Han, S. H. C. (2017). Conceptual review of underrepresentation of women in senior leadership positions from a perspective of gendered social status in the workplace: Implication for HRD research and practice. *Human Resource Development Review, 16*(1), 35–59. doi:https://doi.org/10.1177/1534484317690063.

UNESCO. Director-General 2009–2017. (2017). *Cracking the code: girls' and women's education in science, technology, engineering and mathematics (STEM).* Paris, France: UNESCO.

Wang, M. T., Degol J. L. (2017). Gender gap in science, technology, engineering, and mathematics (STEM): Current knowledge, implications for practice, policy, and future directions. *Educational Psychology Review, 29*(1), 119–140. https://doi.org/10.1007/s10648-015-9355-x.

World Economic Forum. (2016). *The industry gender gap: Women and work in the fourth industrial revolution.* Geneva, Switzerland: World Economic Forum.

Acknowledgments

This publication has been possible with the support of the Erasmus+ Programme of the European Union in its Key Action 2, "Capacity-building in Higher Education". Project W-STEM "Building the future of Latin America: engaging women into STEM" (Reference number 598923-EPP-1-2018-1-ES-EPPKA2-CBHE-JP). The content of this publication does not reflect the official opinion of the European Union. Responsibility for the information and views expressed in the publication lies entirely with the authors.

Contents

Chapter 1
A Model for Bridging the Gender Gap in STEM in Higher Education Institutions

Alicia García-Holgado and Francisco José García-Peñalvo

Abstract Women present a historic and worrying gap in science and technology-related disciplines, generally knowns as STEM (Science, Technology, Engineering, and Mathematics), except in the case of health professions. A holistic approach is needed to support policymakers worldwide in bridging the gender gap in STEM, in which higher education institutions have a crucial role. Promoting this active implication of the universities in this problem, in the European project Building the future of Latin America: engaging women into STEM (W-STEM), a model to modernise the government, management and operation of higher education institutions in Latin America to improve attraction, access to and retention of women in STEM programs has been developed. This situation is not exclusive to Latin American countries, it is a global problem, so the results of the W-STEM project are also applicable to European partners and transferrable worldwide. The main goal of this chapter is to describe the W-STEM model based on three years of working on strategies and mechanisms to improve the attraction, access, guidance, and retention processes to engage more women in STEM programs. The model has been thoroughly tested in eleven institutions in Chile, Colombia, Costa Rica, Ecuador, and Mexico, involving public and private institutions with different gender equality situations.

Keywords Gender gap · Higher education · Gender equality plan · Self-assessment · Women · STEM

A. García-Holgado (✉) · F. J. García-Peñalvo
GRIAL Research Group, Computer Science Department, University of Salamanca, Salamanca, Spain
e-mail: aliciagh@usal.es

F. J. García-Peñalvo
e-mail: fgarcia@usal.es

F. J. García-Peñalvo et al. (eds.), *Women in STEM in Higher Education*, Lecture Notes in Educational Technology, https://doi.org/10.1007/978-981-19-1552-9_1

1.1   Introduction

Gender equality refers to the equal rights, responsibilities and opportunities of women and men and girls and boys. This means that women's and men's rights, responsibilities and opportunities will not depend on whether they are born male or female (OSAGI, 2021). It is goal 5 in the 2030 agenda for Sustainable Development, but also gender equality cuts across all 17 Sustainable Development Goals (SDG) and is reflected in 45 targets and 54 indicators for the SDGs (Dugarova, 2019; Inter-Agency and Expert Group on SDG Indicators, 2017; Pradhan et al., 2017). Gender equality is also part of the European Union (EU) policies so that gender equality and women's empowerment are promoted and financially supported worldwide.

Higher Education Institutions (HEI) have a major role in contributing to the SDGs, not only in their internal policies but also in preparing students for the challenges of the twenty-first century. HEIs are a key element to ensure the sustainability of the SDGs. Besides goal 4, ensure inclusive and equitable quality education and promote lifelong learning opportunities for all, HEIs define processes and actions related to poverty (SDG 1), health and well-being (SDG 3), gender equality (SDG 5) or climate change (SDG13), among others.

According to the Global Gender Gap Report, no country has yet to achieve full gender parity. The Global Gender gap score in 2021 is 67.7%, which means that the remaining gap to close stands at 32.3% (World Economic Forum, 2021). This index measures the gender-based differences in four key dimensions and tracks the narrowing of these gaps over time: economic participation and opportunity, educational attainment, health and survival and political empowerment. Each subindex provides a score related to gender parity. In particular, 95% of educational attainment gaps have been closed already. However, there are gender gaps in higher education worldwide, although the enrolment rates in tertiary education achieved full gender parity in most of the countries (UIS. Stat, 2016).

Women are persistently underrepresented in science, technology, engineering, and mathematics (STEM) (Directorate-General for Research and Innovation Horizon, 2020; OECD, 2015; Tomassini, 2021; UNESCO, 2007; UNESCO Institute for Statistics, 2018). Although most countries have more women than men enrolled in tertiary education, the number of women in tertiary education who choose STEM is around 15% (UNESCO. Director-General 2009–2017, 2017). For example, only 13.76% of women in tertiary education choose STEM compared to 35.12% of men in Colombia (0.39 gender gap score). The situation is worst in Spain, with a gender gap score of 0.33, Finland with 0.25 or Ireland with 0.38.

The STEM labour force is a fundamental tool for responding to the needs of twenty-first- century society; it has a crucial role in sustainable development. Moreover, market forces are transforming industries (World Economic Forum, 2016), not only in favour of technology skills development, but also the need for STEM skills such as critical thinking, problem-solving or innovation (World Economic Forum, 2020). Besides the lack of women in STEM, there is also a lack of workers to fill the needs of the STEM labour force.

On the other hand, evidence shows that increasing gender diversity in STEM may lead to more effective problem-solving and improved innovations (Kahn & Ginther, 2017), and can have long-term effects not only on gender equality but also on economic development (World Economic Forum, 2017). According to Quirós et al. (2018), more women in digital jobs would benefit the European Gross Domestic Product (GDP) by up to 16 trillion euros per year in the European context.

Increasing diversity in STEM, with a particular focus on women, is on the agendas of governments and public and private entities. Foster social inclusion and increasing the female participation in STEM is one of the key challenges of the European Union (García-Holgado et al., 2020a). The EU support actions through initiatives such as the European Platform of Women Scientists (EPWS) or funding projects like INGDVIS (Increasing Gender Diversity in STEM) (Ballatore et al., 2020) or Coding4Girls (Hoić-Božić et al., 2020).

A holistic approach is needed to support policymakers worldwide in bridging the gender gap in STEM. Many factors contribute to this problem, including self-perception, self-efficacy, interest in science, expectations of results, previous educational experiences, family and social context (Lent et al., 1994). According to the SAGA project (STEM and Gender Advancement) (UNESCO, 2016, 2018), the gender gap in STEM should be faced through seven macro objectives covering not only social norms but also educational and professional pathways, and research and decision-making (Fig. 1.1):

- Change perception, attitudes, behaviours, social norms, and stereotypes towards women in STEM in society.

Fig. 1.1 SAGA gender objectives. Based on (UNESCO, 2018)

- Engage girls and young women in STEM primary and secondary education, as well as in technical and vocational education and training.
- Attraction, access to and retention of women in STEM higher education at all levels.
- Gender equality in career progression for scientists and engineers (S&E).
- Promote the gender dimension in research content, practice and agendas.
- Promote gender equality in STEM-related policymaking.
- Promote gender equality in science and technology-based entrepreneurship and innovation activities.

In this holistic approach, higher education institutions have a major role. They directly impact primary and secondary education because they prepare future teachers, so HEI can work on training pre-service teachers for fostering STEM vocations with an emphasis on engaging girls and young women. Likewise, HEI should work on measures to ensure gender equality in career development for staff. Furthermore, HEI plays an important role in research, so they can implement strategies and mechanisms to ensure the gender dimension in research content and develop studies that reduce the gender gap in STEM in a feedback loop. On the other hand, HEIs indirectly impact policies and entrepreneurship and innovation activities because they prepare future professionals and decision-makers.

In this context, the European project Building the future of Latin America: engaging women into STEM (García-Holgado et al., 2019; García-Peñalvo, 2019; García-Peñalvo et al., 2019) has developed a model to modernise the government, management and operation of higher education institutions in Latin America to improve attraction, access to and retention of women in STEM programs.

This chapter describes the W-STEM model for bridging the gender gap in STEM in higher education institutions. The chapter has been organised in the following way. Section 2 introduces the W-STEM project. Section 3 presents the W-STEM model. Section 4 describes the main results after piloting the W-STEM model in 10 Latin American universities. Finally, the last section summarises the main conclusions.

1.2   The Project

There is a lack of contribution of Higher Education Institutions to face the challenges related to reducing the gender gap in STEM, as most actions remain at the public policy level. HEIs need to face specific demands:

- Access to data and better analytical methods to establish the real dimension of gender inequality.
- To define the problem scope determining the real margin of influence and action to improve access and participation of Women in STEM.
- To involve leadership and management in mid and long-term strategies ensuring intended actions and resources to make them sustainable.

- To map in a clear way the processes and mechanisms that are potentially blocking and bias in attraction, access, retention, guidance of Women in STEM.
- To develop effective tools to achieve an increase in enrolment rates.
- To improve policies to attract, enrol, support, guide and monitor students in a differentiated manner.

W-STEM project, "Building the future of Latin America: engaging women into STEM" (Reference 598,923-EPP-1–2018-1-EN-EPPKA2-CBHE-JP), is a European project funded by the European Union through the Erasmus + program, capacity-building in Higher Education call. The project is a structural project that seeks a systemic impact in the Latin American region by promoting reforms in higher education systems, modernising policies, governance and strengthening relations between higher education systems and the economic and social environment (García-Holgado et al., 2019).

The project provides strategic intervention affecting gender equality policies, with a special focus on the attraction and guidance of women in STEM careers, important for current and future society. While some European countries are at higher developmental stages, culture on gender equality is embedded in the universities. Thus, they have mature procedures, experiences, regulations, etc., which can be transferred to Latin American institutions through this project. On the other hand, the gender gap is a global problem, so the European institutions will also learn how to improve their processes.

The funding period started in January 2019 and will finish in July 2022, although the network will continue reducing the gender gap in STEM (https://wstemproj ect.eu/). The consortium consists of fifteen HEI, five from Europe and ten from Latin America:

- University of Salamanca – USAL (Spain) as coordinator.
- Universidad del Norte – UNINORTE (Colombia).
- Oulu University – OULU (Finland).
- Politecnico di Torino – POLITO (Italy).
- Technological University Dublin – TUD (Ireland).
- Northern Regional College – NRC (United Kingdom).
- Tecnológico de Monterrey – ITESM (Mexico).
- Universidad de Guadalajara – UDG (Mexico).
- Universidad Técnica Federico Santa María – UTSM (Chile).
- Universidad Pontificia Católica de Valparaíso – PUCV (Chile).
- Universidad Tecnológica de Bolívar – UTB (Colombia).
- Tecnológico de Costa Rica – ITCR (Costa Rica).
- Universidad de Costa Rica – UCR (Costa Rica).
- Universidad Técnica Particular de Loja – UTPL (Ecuador).
- P15 Technical University of the North – UTN (Ecuador).

Furthermore, each university has also involved secondary education schools as associated partners to work with them in the attraction processes of girls and young women to STEM studies. Finally, UNESCO also participates as associated partner and Columbus Association as external evaluator.

1.3 W-STEM Model

Higher Education Institutions can directly impact the attraction, access to and retention of women in STEM higher education at all levels (UNESCO, 2016, 2018). First, the attraction processes can impact before the students get to university. Later, the access processes mainly affect the application and enrolment activities when women students try to join STEM programs.

There is a critical issue regarding guidance and retention when women are at university due to dropout rates, both while studying and when they join the labour market. Previous studies identify the support received by the STEM students from their academic institutions and teachers is low (García-Holgado et al., 2020; Peña-Calvo et al., 2016). For this reason, HEIs must improve these processes.

The W-STEM model proposes a workflow with different tools and guidelines to impact these three processes, attraction, access and guidance/retention from the HEIs. The workflow involves four phases (Fig. 1.2). The first phase, situation analysis, is review and self-assessment tasks focused on reflection and insight. It is necessary to identify the current state of women in STEM programs because the rate of women in each program depends on different factors; it is not the same for science, technology, engineering or mathematics-related degrees.

The second phase starts once we know the real situation of our institution. This phase focuses on defining the Gender Equality Action Plan (GEAP) to define the

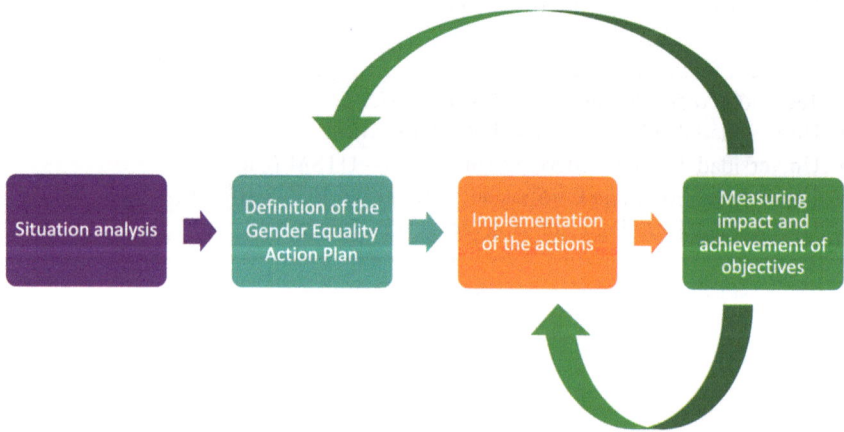

Fig. 1.2 W-STEM workflow to reduce the gender gap in STEM

strategy and the objectives, and design actions to achieve them. Despite Gender Equality Plans (GEP) being mandatory at the university level in most European countries, this is not always required in Latin American HEIs. Moreover, the GEP usually focuses on gender equality among the staff and academic community but not are directly related to reducing the gender gap in STEM.

The next phase covers implementing the actions defined in the Gender Equality Action Plan. These actions work on attraction, access, and guidance processes. Finally, the last phase ensures compliance with the objectives established in the GEAP. It implements mechanisms to measure the impact and achievement of the objectives, and the results are used to update the GEAP or improve the implementation of the actions.

1.3.1   Situation Analysis

The main objective of the first phase is to identify the current situation of women in STEM programs inside the HEI. The results associated with the situation analysis are three:

- Programs with a significant gender gap.
- Knowledge about the processes inside the institution regarding attraction, access to and guidance.
- Good practices inside and outside the institution.

This phase is also divided into three stages or sub-phases, each one with an instrument to collect the required information and a dataset that serves as input for the definition of the gender equality action plan in STEM programs. Moreover, each stage is focused on one of the mentioned results and is an input for the next stage. Figure 1.3 summarises the process.

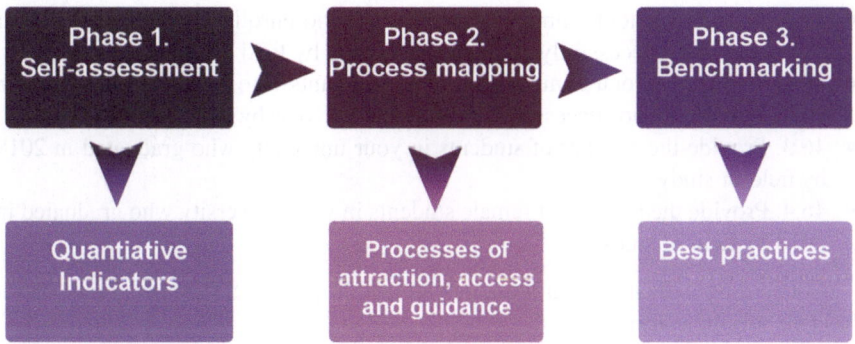

Fig. 1.3 Situation analysis stages

1.3.1.1   Self-Assessment

The first stage is "self-assessment". It focuses on measuring STEM gender equality in the university concerning attraction, access, guidance, retention and other processes, only on the quantitative side. The data collected is key for the rest of the process because it will determine the strategy defined in phase 2, the Gender Equality Action Plan.

We designed a survey to collect quantitative indicators related to undergraduate education level only (bachelor's degree or equivalent), although it is possible to use it for a master's degree or doctorate. The baseline of this instrument is the UNESCO SAGA Toolkit, which is a conceptual and methodological framework to provide a series of tools to integrate, monitor and evaluate gender equality in STEM and assist in the design of gender-sensitive and evidence-based policies to strengthen the gender policy agenda (UNESCO, 2017). The survey is made up of items from the SAGA Indicator Matrix, a tool with 45 indicators to assess on an institutional level the effects of policies and instruments in science, technology and innovation as well as availability and possible need to further develop such policies and instruments to advancing women's participation and career advancement in STEM fields (UNESCO, 2017).

The indicators are organised into ten categories: institutional background information (total number of students and staff: males/females); STEM Programs according to ISCED 2013 classification—broad field; students; attraction; access; enrolment; discrimination; sexual harassment; guidance; dropouts.

The W-STEM self-assessment survey includes indicators from 4 to 26. Moreover, we modified indicator 9 "Total and share of women graduated from university programs by field of study and by educational level," replaced by "Total and share of women graduated from university programs by field of study". Besides, the survey includes two new indicators, 46 and 47. Indicator 46 is based on indicator 9, but focused on guidance of women enrolled and graduated in STEM programs, which also consists of four sub-indicators:

- 46.1. Provide the total number of applicants who enrolled in first year in your university who successfully completed first year by field of study.
- 46.2. Provide the total number of female applicants who enrolled in first year in your university who successfully completed first year by field of study.
- 46.3. Provide the number of students in your university who graduated in 2018 by field of study.
- 46.4. Provide the number of female students in your university who graduated in 2018 by field of study.

Concerning indicator 47, it measures female dropout in STEM programs, and it has two sub-indicators:

- 47.1. Total dropout at first year.
- 47.2. Share of the female dropouts during the first year.

Finally, some indicators are complemented with a set of questions to identify related policies. The survey template is available on (W-STEM Consortium, 2019a). The self-assessment must be applied at the beginning of the process, collecting information from the last academic year. Although covering more academic years is possible, one year is enough due to a challenge to feed the indicators with sufficient data (García-Holgado et al., 2020c). On the other hand, all indicators are important; however, not all are mandatory. At least the institution should complete the indicators, 4 to 8, 14, 15, 46 and 47.

1.3.1.2 Process Mapping

Higher Education Institutions are complex organisations due to the multiplicity of purposes, the involvement of multiple actors in the decision-making processes, the different organisational cultures that coexist within the organisation or the complex structure that derives from its nature. Even though the people working on the gender gap in STEM is part of the institution, most will not be aware of all the details of the institution's functioning unless they have been part of the management team.

This stage provides a clear map of all the steps and the involved stakeholders in the key processes of the project: attraction, access, retention, and guidance. It complements the results of the indicators found during the self-assessment from a qualitative point of view and allows identifying bottlenecks, deficiencies, possibilities and who are involved and their roles.

The definition of the process mapping template (W-STEM Consortium, 2019b) and the guidelines was supported by an external expert to ensure that data collection is carried out in a similar way regardless of the institution and that all necessary elements are covered. The mapping should include plans, strategies and policies, measures and initiatives, institutional framework at different levels (university, department, etc.), and any stage (planning/execution/evaluation or closing phases, in development, or already finished).

The template has four datasheets, one per macro-categories (attraction, access, guidance and retention). Each sheet provides a set of columns to organise the input information. Each macro category is divided into subcategories and activities. For example, the access category could be divided into admission to the first undergraduate semester, tuition payment, academic enrolment, etc. These subcategories will depend on the institution.

Moreover, each subcategory is divided into activities and the following information is provided: the name of the activity, a short description of the activity and the impact on the target groups. Finally, the departments, services or units in charge of each subcategory and/or activity should be provided (García-Holgado et al., 2020c).

The mapping process can be carried out after the self-assessment or parallel and should reflect the current situation. It is a snapshot of the institution's performance on gender gap issues with special attention to STEM. The output of this stage will be useful to involve decision-makers in the next phase, the definition of the Gender Equality Action Plan.

1.3.1.3 Benchmarking

The gender gap in STEM is not a new issue; many organisations and communities are working on reducing the gender gap in STEM. Initiatives such as Niñas Pro in Chile (Vidal et al., 2021), a Female Engineer in Every School in Spain (Cerezo et al., 2018) or Women in Science & Engineering (WiSE) in the United States (Paderewski-Rodríguez et al., 2017) are some examples of actions developed by Higher Education Institutions to engage more girls into STEM.

The situation analysis phase includes a stage to consider those actions that already work and learn from the previous problems and decisions. The W-STEM model proposes two ways to conduct this stage. On the one hand, a systematic research projects review (SRPR) (García-Holgado et al., 2020d) to identify funding projects related to the gender gap in STEM. For example, in the European Union, there is a strong investment in projects focused on gender in STEM (García-Holgado et al., 2020a). Those projects have outcomes that can be transferred to other institutions, such as educational materials, online courses, examples of good practices or software tools.

On the other hand, benchmarking rounds focused on attraction, access and guidance/retention. The methodology proposed to conduct the benchmarking round is based on the Columbus Association methodology. It is a process applied in companies and can be transferred to universities. It is a structured process—a series of actions, steps, functions or activities—that leads to comparing services-activities-processes-outputs-outcomes to identify and adopt good practices to improve university performance. In the W-STEM model, benchmarking rounds aim to identify policies, procedures and mechanisms that are considered good practices in attracting, accessing and mentoring women in STEM programs to be adapted to the needs identified in the institution.

The benchmarking can be conducted internally in large universities, for example, with different campuses, but it can also involve several institutions participating in the preparation process and the dynamics afterwards. Each institution has to collect examples of good practices that might be interesting and of novelty from an institutional point of view. This could be related to any of the three processes (attraction, access and guidance or retention); it may even impact more than one process at a time. Furthermore, the institution can identify lessons learned or problems encountered in implementing an action. The preparation process can follow the template prepared in W-STEM (W-STEM Consortium, 2019c). After collecting the information, each institution shares the good practices, lessons learned and problems in an event, face-to-face or online.

1.3.2 Definition of the Gender Equality Action Plan

Typically, a gender equality plan (GEP implies establishing and making effective a set of basic principles to safeguard gender equality in all the spheres of academic life. It also establishes the core actions to be developed in the university, across its different levels: human resources practices and management processes, student services and institutional communication, research design and delivery, and institutional communication (Barros et al., 2018). However, the GEAP proposed in the W-STEM model is supplementary to GEP already developed in some HEI. Despite the GEAP implementing some measures to ensure gender equality inside STEM programs, the principal focus is reducing the gender gap in those programs through gender equality actions. Moreover, the GEAP works in attraction and access, which are not directly internal processes of the institution itself but seek to work outside the institution.

The GEAP includes specific actions to increase access, attraction and guidance of women in STEM programs. The results from the previous phase, situation analysis, are inputs to define the GEAP. First, indicators from the self-assessment will serve to identify the target STEM programs. Second, the process mapping will provide useful information about the processes already implemented in the institution, so the actions defined in the GEAP could change those processes or introduce new ones. Finally, the benchmarking results provide ideas and learning lessons to define some actions to achieve the GEAP objectives.

The definition of the action plan should follow three steps: define the strategy; design the objectives; and define the actions to achieve the objectives. The strategy definition should include a short text to contextualise the action plan. This is not a state of the art or theoretical introduction; its goal is to explain the plan's purpose.

Regarding the actions, it should be noted that each objective must be associated with more than one action; in another case, the objective is not well defined. The actions definition is not only the presentation of the action, but also the process to identify the person or persons responsible for each action, the services in the university that will help to implement the actions and achieve the objectives, and the persons that will support the implementation of the action. Furthermore, the description of the actions will include a timeline to establish the implementation timing for each action and identify the milestones.

1.3.3   Implementation of Actions and Measuring Impact and Achievement of Objectives

The implementation of the actions defined in the GEAP, and monitoring are the last phases of the W-STEM model. The implementation phase depends on the actions defined by the institutions. However, it is necessary to consider the monitoring phase when we develop those actions. For this reason, the actions should include mechanisms to measure the impact to assess the achievement of the GEAP objectives and the quality of the implementation process.

Actions must include surveys, observation mechanisms, evidence gathering, etc., to have enough data to support decision-making processes during the monitoring phase. Furthermore, the measures will have an impact on the actions. Therefore, actions that are carried out on a regular basis will be updated or replaced by others if the expected impact is not achieved.

Finally, it is possible using benchmarking rounds as monitoring activities. These rounds can be organised internally, involving the stakeholders involved in the GEAP, to review the progress done, learn from the different actions implemented, and make decisions about the GEAP. Also, it is possible to invite other institutions to the benchmarking round, not only for identifying good practices, but also for sharing learning lessons and solving common problems.

1.4 Main Results

The results of the W-STEM models depend on each institution. However, this section summarises some relevant results obtained during the implementation and validation of the model in the Latin American institutions involved in the W-STEM project.

In addition to the resources and materials developed to support the implementation of the W-STEM model, there are four primary results associated with the process that can be used in other contexts:

- International Leadership Summer Camp.
- Women in STEM mobile app.
- Attraction campaigns.
- W-STEM Mentoring Network.

1.4.1 International Leadership Summer Camp

First, the International Leadership Summer Camp (also named International Leadership Summit) was organised face-to-face in Cartagena de Indias and Barranquilla (Colombia). It was an event on the global and regional outlook on gender equality at STEM education and women in science participation. The aim was to make the problem visible to the policymakers from the higher education institutions involved in W-STEM to integrate them in the definition of the GEAP as a strategy to ensure the sustainability of the plan. Each participant university involved at least one institutional leader (rectors, vice-rectors, deans of STEM programs, institutional decision-makers). Also, five experts from UNESCO, NRC (UK) and Liquid Galaxy Lab at FAÇENS (Brazil) participated in the event.

The Leadership Summit included sessions for building bridges between national/regional policies and the institutional level. Leaders, experts and the W-STEM team participated in a World Café, a set of roundtables focused on building bridges between public policy and institutional initiatives concerning the mechanisms and strategies about attraction, access and guidance of women in STEM programs (García-Peñalvo et al., 2019).

The second part of the Leadership Summit was focused on the first benchmarking round and a workshop to start the definition of the Gender Equality Action Plans.

1.4.2 Women in STEM Mobile App

The purpose of the W-STEM app is to give visibility to women in STEM careers, not only women of a very high level in STEM, but also young women with different profiles—last year students, PhD students, young researchers, developers, etc. The app is a tool for schools and high schools in Latin America to show young women in STEM since the books and materials used in the classroom only show examples of men. It is intended to give young people ideas about what it means or why to select a STEM program (García-Holgado et al., 2020e). The app is available for iOS and Android, both for smartphones and tablets (https://wstemproject.eu/app/).

The app has two types of content. On the one hand, short bios provide some useful information to potential STEM students and share with them similar concerns that other people have had (Marín-Raventós et al., 2020). We used a survey based on (Ballatore et al., 2019) to identify prospective students, successful graduates and current students, with key aptitudes correlated to STEM disciplines/professions. A total of 6358 profiles were collected, 2071 have finished their university studies, and 4287 are students (45.33% male, 54.03% female, 0.34% preferred not to answer, and 0.3% non-binary).

On the other hand, the app facilitates access to 335 interviews of women in STEM from different countries, cultural backgrounds, ages, and career stages (not only senior positions). The goal was to show diversity in STEM and avoid stereotyping. The interviews duration is between 3 and 7 min, and the original language is Spanish or English with subtitles. Figure 1.4 shows a sample of the videos available on YouTube (W-STEM Consortium, 2020).

1.4.3 Attraction Campaigns

The original plan was to conduct the attraction campaigns inside the secondary schools to engage the young women students and support them to enrol in a STEM program. However, the COVID-19 crisis impacted on education (García-Peñalvo, 2021; García-Peñalvo et al., 2020, 2021; Knopik & Oszwa, 2021) and forced the transformation of the attraction campaigns into virtual campaigns.

The involvement of the schools during the lockdown period in 2020 hindered the organisation of some activities. For this reason, the attraction campaigns also involved awareness-raising campaigns. Moreover, all the events were online, so a common schedule was prepared to avoid overlapping. Although there were many activities, the attraction campaigns guidelines included the following type of events:

Fig. 1.4 Women in STEM interviews available on the W-STEM YouTube channel

- Awareness webinars: the objective is to raise awareness among schoolgirls about the gender gap in STEM fields.
- Informative webinars: the objective is to provide relevant information about the different STEM areas, their fields of action, their importance for society, etc.
- Webinars aimed at training in STEM knowledge: applications of the STEM areas that are compelling and that can be taught to girls while addressing real problems. For example, data analysis, programming principles.
- Activities with professional women from different countries: such as Q&A sessions (live questions from participants) or panels/talks about the differences between careers and their fields of application.
- Talks with guests on the experience of studying and/or practicing STEM, as well as the role that women have had in the history of that field.
- Coffee with a STEM woman: more informal conversations with a guest.
- Virtual Poster Fair: exhibitions of women who have contributed to STEM areas with Q&A.
- Film forum: girls are encouraged to watch a film that addresses the gender gap in STEM or highlights a STEM woman's story and then virtual discussion space is opened.
- Vocational orientation: smaller spaces where a specific career is addressed. In these activities, the idea is to explain the most relevant aspects of each STEM career but separately and in smaller groups so that the doubts and questions of most of the attendees can be answered.

1.4.4   W-STEM Mentoring Network

We have created a network of mentoring programs with pilot experiences the eleven HEI from Ibero-America (Chile, Colombia, Costa Rica, Ecuador, Mexico, and Spain).

The Mentoring Network aims to empower women and encourage their active participation in STEM programs. The mentoring model implemented uses a three-way relationship. It involves a mentor teacher, the peer-mentor and the mentee (González Rogado et al., 2021). The guidelines for implementing the mentoring programs are common for all the network, but each institution has adapted them to its own contexts and needs according to its GEAP.

First, mentor teachers and peer-mentors receive cross-cutting training in leadership, women's empowerment, inclusive language and creating inclusive environments. Depending on the institution, mentor teachers and peer-mentors are men and women or only women. Regarding the mentees, they are first-year STEM female students. The goal is to support and empower them during their first year of studies.

1.5   Conclusions

The gender gap in STEM is a reality that affects to a different extent by countries. Although there are countries with less gender gap in STEM, such as Oman, Myanmar, Morocco, Mauritania, Gambia or Benin, most countries have a remaining gap to close stands at 60–70% (World Economic Forum, 2021).

Higher Education Institutions have a major role in reducing the gender gap in STEM because they impact before, during and after tertiary studies. For this reason, HEI must implement gender equality action plans based on their gender equality situation. This chapter presents the W-STEM model, a proposal based on three years working on strategies and mechanisms to improve the attraction, access, guidance, and retention processes to engage more women into STEM programs.

The W-STEM model has been fully tested in eleven institutions in Chile, Colombia, Costa Rica, Ecuador and Mexico, involving public and private institutions with different gender equality situations. Each institution has adapted the model and the resources to develop the different context.

However, several caveats need to be noted regarding the W-STEM model. Although work has been done on the three processes of attraction, access, guidance and retention, the actions associated with access are lower than others. The main problem is that national education laws and policies regulate the access processes, so the institution could influence the support provided to the students during those processes but not in the process itself.

On the other hand, concerning the situation analysis, some important limitations need to be considered. First, the self-assessment and mapping process require the collaboration of different services and units, so there is a high risk of not achieving all information. HEI does not have a unique service that centralises statistical information.

Finally, there is a definite need to transfer the W-STEM model to other HEI, with a particular focus in Latin America. For this reason, the W-STEM project is developing a set of online training modules on policies, mechanisms and processes, as well as on the implementation of the W-STEM model to enhance the attraction, access and guidance of women in STEM.

Acknowledgements This work has been possible with the support of the Erasmus+ Programme of the European Union in its Key Action 2, "Capacity-building in Higher Education". Project W-STEM "Building the future of Latin America: engaging women into STEM" (Reference number 598923-EPP-1-2018-1-ES-EPPKA2-CBHE-JP). The content of this publication does not reflect the official opinion of the European Union. Responsibility for the information and views expressed in the publication lies entirely with the authors.

References

Ballatore, M. G., Borger, J. D., Misiewicz, J., & Tabacco, A. (2020). ANNA tool: A way to connect future and past students in STEM. *IEEE Revista Iberoamericana De Tecnologias Del Aprendizaje, 15*(4), 344–351. https://doi.org/10.1109/RITA.2020.3033231.

Ballatore, M. G., Barman, L., Borger, J. D., Ehlermann, J., Fryers, R., Kelly, K., Misiewicz, J., Naimi-Akbar, I., & Tabacco, A. (2019). Increasing gender diversity in STEM: A tool for raising awareness of the engineering profession. In M. Á. Conde-González, F. J. Rodríguez Sedano, C. Fernández Llamas, & F. J. García-Peñalvo (Eds.), *Proceedings of the 7th International Conference on Technological Ecosystems for Enhancing Multiculturality (TEEM 2019)* (León, Spain, October 16–18, 2019). ACM, New York, NY, USA, pp. 216–222. https://doi.org/10.1145/3362789.336 2832.

Barros, V. F. A., Vasconcelos, R. M., Araújo, E., Amaral, L., & Ramos, I. (2018) A positive perspective to implementation of a gender equality plan: A question of design, time and participation. In *2018 IEEE Frontiers in Education Conference (FIE)*, pp. 1–5. https://doi.org/10.1109/FIE.2018.8659112.

Cerezo, E., Ayuso, N., Trillo, R., Masiá, B., Murillo, A. C., Mariscal, L., Ruberte, L., Baldassarri, S., Villarroya, M., Delgado, M., & Mayoral, C. (2018). A female engineer in every school. In C. Manresa-Yee & R. Mas Sansó (Eds.), *Proceedings of the XIX International Conference on Human Computer Interaction*, ACM, New York, NY, USA. p. Article 38. https://doi.org/10.1145/3233824.3233847.

Directorate-General for Research and Innovation Horizon. (2020). Science with and for Society (2021) She figures-gender in research and innovation statistics and indicators. *European Commission, Brussels*. https://doi.org/10.2777/06090.

Dugarova, E. (2019). Gender equality as an accelerator for achieving the sustainable develpment goals.

García-Holgado, A., Camacho Díaz, A., & García-Peñalvo, F. J. (2019). Engaging women into STEM in Latin America: W-STEM project. In M. Á. Conde-González, F. J. Rodríguez Sedano, C. Fernández Llamas, F. J. García-Peñalvo (Eds.), *Proceedings of the 7th International Conference on Technological Ecosystems for Enhancing Multiculturality (TEEM 2019)* (León, Spain, October 16–18, 2019). ACM International Conference Proceeding Series (ICPS). ACM, New York, NY, USA, pp. 232–239. https://doi.org/10.1145/3362789.3362902.

García-Holgado, A., Verdugo-Castro, S., González, C. S., Sánchez-Gómez, M. C., & García-Peñalvo, F. J. (2020a). European proposals to work in the gender gap in STEM: A systematic analysis. *IEEE Revista Iberoamericana De Tecnologías Del Aprendizaje, 15*(3), 215–224. https://doi.org/10.1109/RITA.2020.3008138.

García-Holgado, A., González-González, C. S., & Peixoto, A. (2020b). A comparative study on the support in engineering courses: A case study in Brazil and Spain. *IEEE Access, 8*, 125179–125190. https://doi.org/10.1109/ACCESS.2020.3007711.

García-Holgado, A., Mena, J., García-Peñalvo, F. J., Pascual, J., Heikkinen, M., Harmoinen, S., García-Ramos, L., Peñabaena-Niebles, R., & Amores, L. (2020c). Gender equality in STEM programs: A proposal to analyse the situation of a university about the gender gap. In *2020 IEEE Global Engineering Education Conference (EDUCON)*, (27–30 April 2020, Porto, Portugal). IEEE, USA, pp. 1824–1830. https://doi.org/10.1109/EDUCON45650.2020c.9125326.

García-Holgado, A., Marcos-Pablos, S., & García-Peñalvo, F. J. (2020d). Guidelines for performing systematic research projects reviews. *International Journal of Interactive Multimedia and Artificial Intelligence, 6*(2), 137–144. https://doi.org/10.9781/ijimai.2020.05.005.

García-Holgado, A., Verdugo-Castro, S., Sánchez Gómez, M. C., & García-Peñalvo, F. J. (2020e). Facilitating access to the role models of women in STEM: W-STEM mobile app. In P. Zaphiris & A. Ioannou (Eds.), *Learning and Collaboration Technologies. Designing, Developing and Deploying Learning Experiences*. HCII 2020e. Lecture Notes in Computer Science, vol 12205. Springer, Cham, pp. 466–476. https://doi.org/10.1007/978-3-030-50513-4_35.

García-Peñalvo, F. J. (2019). Women and STEM disciplines in Latin America. The W-STEM European Project. *Journal of Information Technology Research* 12(4), v–viii.

García-Peñalvo, F. J. (2021). Digital transformation in the universities: Implications of the COVID-19 pandemic. Education in the Knowledge Society 22.

García-Peñalvo, F. J., Bello, A., Domínguez, A., & Romero Chacón, R. M. (2019). Gender balance actions, policies and strategies for STEM: Results from a world café conversation. *Education in the Knowledge Society, 20*(15). https://doi.org/10.14201/eks2019_20_a31.

García-Peñalvo, F. J., Corell, A., Abella-García, V., & Grande-de-Prado, M. (2020). Online assessment in higher education in the time of COVID-19 Education in the Knowledge Society 21. https://doi.org/10.14201/eks.23086.

García-Peñalvo, F. J., Corell, A., Abella-García, V., & Grande-de-Prado, M. (2021). Recommendations for man-datory online assessment in higher education dur-ing the COVID-19 pandemic. In D. Burgos, A. Tlili, & A. Tabacco (Eds.), *Radical solutions for education in a crisis context. COVID-19 as an opportunity for global learning*. Singapore: Springer Nature, pp. 85–98.

González Rogado, A. B., García-Holgado, A., & García-Peñalvo, F. J. (2021). Mentoring for future female engineers: Pilot at the Higher Polytechnic School of Zamora. In A. García-Holgado, F. J. García-Peñalvo, C. S. González González, A. Infante Moro, & J. C. Infante Moro (Eds.), *2021 XI International Conference on Virtual Campus (JICV)*. IEEE, USA. https://doi.org/10.1109/JICV53222.2021.9600410.

Hoić-Božić , N., Holenko Dlab, M., & Franković, I., & Ivašić-Kos, M. (2020). Teaching programming skills to girls. In M Baptista Nunes & P Isaias (Eds.), *Proceedings of the IADIS International Conference e-Learning 2020* (pp. 153–156). IADIS.

Inter-Agency and Expert Group on SDG Indicators. (2017). Global indicator framework for the Sustainable Development Goals and targets of the 2030 Agenda for Sustainable Development. United Nations.

Kahn, S., & Ginther, D. (2017). *Women and STEM*. National Bureau of Economic Research.

Knopik, T., & Oszwa, U. (2021). E-cooperative problem solv-ing as a strategy for learning mathematics during the COVID-19 pandemic. Education in the Knowledge Society 22.

Lent, R. W., Brown, S. D., & Hackett, G. (1994). Toward a unifying social cognitive theory of career and academic interest, choice, and performance. *Journal of Vocational Behavior, 45*(1), 79–122. https://doi.org/10.1006/jvbe.1994.1027.

Marín-Raventós, G., Romero, R. M., & Monge-Soto, A. L. (2020). Using student profiles to motivate and understand how to attract women to computer science. In G. Rodríguez-Morales, A. García-Holgado (Eds) Proceedings of the XII Latin American Women in Computing Congress 2020 (LAWCC 2020), Loja, Ecuador, October 19, 2020. vol CEUR Workshop Proceedings. CEUR-WS.org, pp 1–12.

OECD. (2015). The ABC of gender equality in education. *OECD Publishing, Paris*. https://doi.org/10.1787/9789264229945-en.

OSAGI (Office of the Special Advisor on Gender Issues and Advancement of Women). (2021). Gender mainstreaming: Strategy for promoting gender equality. Fact sheet.

Paderewski-Rodríguez, P., García-Arenas, M. I., Gil-Iranzo, R. M., González, C. S., Ortigosa, E. M., & Padilla-Zea, N. (2017). Initiatives and strategies to encourage women into engineering. *IEEE Revista Iberoamericana De Tecnologias Del Aprendizaje, 12*(2), 106–114. https://doi.org/10.1109/RITA.2017.2698719.

Peña-Calvo, J.-V., Inda-Caro, M., Rodríguez-Menéndez, C., & Fernández-García, C.-M. (2016). Perceived supports and barriers for career development for second-year STEM students. *Journal of Engineering Education, 105*(2), 341–365. https://doi.org/10.1002/jee.20115.

Pradhan, P., Costa, L., Rybski, D., Lucht, W., & Kropp, J. P. (2017). A systematic study of sustainable development goal (SDG) interactions. *Earth's Future, 5*(11), 1169–1179. https://doi.org/10.1002/2017EF000632.

Quirós, C. T., Morales, E. G., Pastor, R. R., Carmona, A. F., Ibáñez, M. S., & Herrera, U. M. (2018). Women in the digital age. *Publications Office of the European Union, Luxembourg*. https://doi.org/10.2759/526938.

Tomassini, C. (2021). Gender gaps in science: Systematic review of the main explanations and the research agenda. *Education in the Knowledge Society* 22:Article e25437. https://doi.org/10.14201/eks.25437.

UIS. Stat. (2016). http://data.uis.unesco.org/.

UNESCO. (2007). *Science, technology and gender: An international report*. UNESCO Publishing, Paris, France.

UNESCO. (2016). Measuring gender equality in science and engineering: The SAGA science, technology and innovation gender objectives list (STI GOL). SAGA Working paper 1. Paris, France: UNESCO.

UNESCO. (2017). Measuring gender equality in science and engineering: The SAGA toolkit. SAGA Working Paper 2. Paris, France: UNESCO.

UNESCO. (2018). Telling SAGA: Improving measurement and policies for gender equality in science, technology and innovation. SAGA Working Paper 5. Paris, France: UNESCO.

UNESCO. Director-General 2009–2017. (2017). *Cracking the code: Girls' and women's education in science, technology, engineering and mathematics (STEM)*. UNESCO.

UNESCO Institute for Statistics. (2018). *Women in science*. UNESCO Institute for Statistics.

Vidal, M., Maldonado, J., Bracamonte, T., Miranda, F., Labarca, A., & Simmonds, J. (2021). Niñas Pro: An initiative to educate, inspire and empower women. In M. Estrada & A. García-Holgado (Eds.), *Proceedings of the XIII Congress of Latin American Women in Computing 2021 (LAWCC 2021) Co-located with XLVII Latin American Computer Conference (CLEI 2021)*, San José, Costa Rica, October 28, 2021. vol. CEUR Workshop Proceedings. CEUR-WS.org, pp. 35–46.

World Economic Forum. (2016). *The industry gender gap: Women and work in the fourth industrial revolution*. Geneva, Switzerland: World Economic Forum.

World Economic Forum . (2017). *The Global Gender Gap Report 2017*. Geneva, Switzerland: World Economic Forum.

World Economic Forum. (2020). *The future of jobs Report 2020*. Geneva, Switzerland: World Economic Forum.

World Economic Forum. (2021). *The global gender gap report 2021*. Insight Report. Geneva, Switzerland: World Economic Forum.

W-STEM Consortium. (2019a). W-STEM Self-assessment Matrix. https://doi.org/10.5281/zenodo.3594822.

W-STEM Consortium. (2019b). W-STEM Process Mapping Template. https://doi.org/10.5281/zen odo.3594845.

W-STEM Consortium. (2019c). W-STEM benchmarking questionnaire form. https://doi.org/10.5281/zenodo.3594858.

W-STEM Consortium. (2020). W-STEM: Women in STEM Internviews. https://youtube.com/pla ylist?list=PL43UVswQuVDMrDJJvzbnnKco1CoJJhZMK.

Chapter 2
A Review of Irish National Strategy for Gender Equality in Higher Education 2010–2021

Julie Dunne, Ayesha O'Reilly, Ashley O'Donoghue, and Mary Kinahan

Abstract This chapter provides a narrative of the key policies, initiatives and actions that have transformed both the landscape of gender equality in Higher Education in Ireland, and the role of gender equality in research policy and public engagement in STEM over the last decade. It aims to provide leaders committed to gender equality with examples of good practice within the EU-Irish context. The first part of the chapter focuses on the EU gender equality strategies for Higher Education. It explains the review of gender equality undertaken by the Irish Higher Education Authority (HEA), the significant findings, and the National Gender Action Plan designed to address the issues identified. It contains examples of high-level initiatives implemented to deliver on key actions. These include centres of excellence for Gender Equality, and affirmative actions taken to address imbalance at senior levels in the Irish Higher Education system. It also provides an account of Irish participation in the UK Advance Higher Education (*Advance HE*) gender supports including the women's leadership programme '*Aurora*', and the '*Athena SWAN*' charter that provides an accreditation framework for auditing, supporting and transforming gender equality in Higher Education Institutions. For context, some examples of implementing these in an Irish Institution are provided. The chapter then reviews the gender equality strategies and policies of the main research funding organisations in Ireland, namely Science Foundation Ireland (SFI), and the Irish Research Council (IRC). It indicates how gender equality manifests in achieving gender balance in the Irish research community; and in funding applications and consequently in research design to embed the gender element. It also provides an

J. Dunne (✉) · A. O'Reilly · A. O'Donoghue · M. Kinahan
Technological University Dublin, Dublin, Ireland
e-mail: julie.dunne@tudublin.ie

A. O'Reilly
e-mail: ayesha.oreilly@tudublin.ie

A. O'Donoghue
e-mail: ashley.odonoghue@tudublin.ie

M. Kinahan
e-mail: mary.kinahan@tudublin.ie

© The Author(s) 2022

F. J. García-Peñalvo et al. (eds.), *Women in STEM in Higher Education*, Lecture Notes in Educational Technology, https://doi.org/10.1007/978-981-19-1552-9_2

overview of recent national studies carried out to analyse the Irish public's perceptions and awareness of STEM in society, and factors leading to career and study choices by young Irish people. It signposts to the recent actions to address gender equality provided under SFI's remit for public engagement in STEM.

Keywords Gender equality policy · Ireland · European union (EU) · Higher education · Research funding · Advance HE

2.1 Introduction

Despite equality legislation and equality policies, gender inequality persists in Ireland and globally in many sectors of society. A World Economic Forum report on gender gaps in health, education, work and politics states it will take 99.5 years to close these gaps (World Economic Forum, 2019). In relation to the gender income gap, they found that women are heavily under-represented in technology (12%), engineering (15%), data and AI (26%). These are sectors which have experienced substantial wage increases in the last decade. The European Commission's She Figs. 2021 shows women are under-represented in technical professions with women representing 24.9% of self-employed professionals in Science and Engineering (S&E) and Information & Communication Technologies (ICT). In addition, women are under-represented in the highest academic levels (26.2%) and at the highest decision-making levels (26.2%) in academic institutions. This reflects the global issue of gender inequality in Higher Education research and innovation (Commission, 2021). Gender equality is an important ethical, legal and performance issue for Higher Education. Ethically, the United Nations Sustainable Development Goal 5 calls for Gender Equality and the right for women and girls to have equal participation in all aspects of society. Legally, many countries have equality legislation that prevents gender discrimination. In Ireland for example, the Employment Equality Act 1998 specifically legislates for the prevention of gender discrimination in the fair and equal access to employment and promotion opportunities (Employment Equality Act, 1998). In terms of overall performance, research shows that diversity is a key strength. Gender balance on executive boards is positively correlated with increased performance of organisations (Gratton et al., 2007; Carter & Wagner, 2011; Reinert et al., 2016) and research publications from ethnically diverse research teams are more highly cited than those published by ethnically homogenous teams (Hunt et al., 2015). Diversity brings different perspectives that increases creativity and innovation (Gassmann, 2001; Dai et al., 2019).

European Union Gender Equality Strategy.

Gender equality is a core principle of the European Union, but it is not yet a reality. In business, politics and society as a whole, we can only reach our full potential if we use all of our talent and diversity. Using only half of the population, half of the ideas or half of the energy is not good enough.

President of the European Commission, Ursula von der Leyen.

The European Commission states that 'the under-representation of women threatens the goals of science in achieving excellence, as well as being wasteful and unjust' (European Commission, 2000). Building on the strengths and lessons learned from the Strategic Engagement for Gender Equality 2016–2019 (Publications Office of the European Union, 2020a) and to address the persistent under-representation of women and girls in the labour market, the European Commission (2020b) developed the Gender Equality Strategy 2020–2025.

The European Commission Gender Equality Strategy 2020–2025 has identified the following concrete actions to address the equal participation of women and girls in the economy and society:

– Improve the balance between women and men in decision-making positions, including on company boards and in politics. Adopt EU-wide targets on gender balance on corporate boards. Encourage the participation of women as voters and candidates in the 2024 European Parliament elections. As one of the first deliverables of the Strategy, the Commission will propose binding pay transparency measures. Encourage a more balanced participation of women and men in all work sectors for more diversity in the workplace.
– Promote the EU Platform of Diversity Charters in all sectors.
– Address the digital gender gap in the updated Digital Education Action Plan.
– Gender mainstreaming. Include a gender perspective in all policy areas, at all levels and at all stages of policymaking. For example: Specific needs, challenges and opportunities in different sectors, such as transport, energy and agriculture, will be addressed within the Commission's gender mainstreaming actions.
– The Commission will address the gender dimension in its major initiatives responding to European challenges such as climate change and digitalisation.
– Intersectionality. All women are different and may face discrimination based on several personal characteristics. For instance, a migrant woman with a disability may face discrimination on three grounds. The intersectionality of gender with other grounds of discrimination will be addressed across EU policies.
– Dedicated funding for a gender-equal future. EU budget (2021–2027) will make funding available through a number of EU programmes for gender equality-related projects. For example, dedicated grants under the Citizens, Equality, Rights and Values Programme to the big structural, social and cohesive EU funds.

As part of the W-STEM project (García-Holgado et al., 2019; García-Peñalvo et al., 2019) TU Dublin staff from both professional services and academia have provided an overview of the gender equality policies and actions that have influenced the Irish context over the last decade. In the following sections, (1) the initiatives to support achieving gender equality in Irish Higher Education Institutions are discussed. These include undertakings by the Irish Higher Education Authority (HEA), specifically (1.1) a review of gender equality in Irish HEAs and (1.2) formation of a gender task force. The overview also includes (2) the UK-based

Advance HE and the support it provides to Ireland and elsewhere including (2.1) Athena SWAN and (2.2) Aurora Women's Leadership programme. Thereafter, (3) a discussion follows on gender equality within Irish Research Funding organisations including (3.1) the Irish Research Council and (3.2) Science Foundation Ireland. The chapter closes with some concluding remarks which may be useful to those interested in understanding the systematic approach taken by Ireland and the EU to advance gender equality in Higher Education.

2.2 Gender Equality in Irish Higher Education Institutions

To address the under-representation of women in senior levels across Irish Higher Education Institutions, several initiatives were implemented and are discussed in the following section.

2.2.1 The Higher Education Authority: National Review of Gender Equality in Irish Higher Education Institutions

The Higher Education Authority (HEA) conducted a National Review of Gender Equality in Irish Higher Education Institutions (Higher Education Authority, 2016). This expert review aimed to address the numerous factors within Higher Education Institutions (HEIs), conscious and unconscious, cultural and structural, that result in women facing barriers to progression which are not experienced to the same degree by their male colleagues. The review conducted an in-depth analysis of the gender balance of academic and non-academic (professional service and support staff) staff across all grades of employment as well as institutions' management teams, academic councils and governing boards.

Review Process

The review process began with the development of the Terms of Reference and appointment of the Expert Group. The approach taken in conducting the review involved the following stages:

– Policy context research.
– Literature review of the international and national challenges and emerging solutions.
– Data collection to establish figures on the gender breakdown of HEI staff—identifying gaps in the data gathered and analysis of the data available.

Table 2.1 HEA survey of staff in the Irish Higher Education Sector (n = 4,835): priority areas identified to address barriers to women's progression

Priority Areas
67% Promotion/progression criteria
61% Gender balance on senior management teams at institutional level
60% Overall culture
52% Career development opportunities
51% Transparent procedures/processes
51% Senior management's leadership on gender equality
50% Representation of men and women on key committees
50% Childcare/carers' provision and supports

– Collection of HEI institutional equality policies, and where applicable, HEI Athena SWAN applications or institutional statements on their gender equality initiatives.
– Wide consultation with stakeholders involving face-to-face meetings with the Expert Group, written submissions from interest groups, and a public online survey. See Table 2.1 for priority areas identified.
– Development of recommendations.
– Feedback from survey respondents (Table 2.1) and wider consultations informed the HEA recommendations to address gender inequality in Higher Education.

HEA recommendations were identified for four key stakeholder groups to drive positive change for gender equality:

– Higher education institutions.
– The Higher Education Authority.
– Research funding and related agencies.
– Wider Higher Education stakeholders.

A total of 22 recommendations were specifically identified for *Higher Education Institutions* and, for context, the TU Dublin response to implementing these recommendations is outlined in Table A.1 in the Appendix to this chapter. Key initiatives implemented by TU Dublin include:

– *Structural Changes.* All university committees with responsibility for operating and resourcing decisions must be gender balanced.
– *Cultural Changes.* To create an understanding of how unconscious bias can lead to stereotyping, prejudice and discrimination, unconscious bias training is provided for all staff and included in leadership development, performance management and interview skills training.
– *Policy Changes.* All recruitment and selection panels must be gender balanced and all panel members must complete unconscious bias training.

– *Career Development.* To support career development and progression, the university sponsors 20 women annually to attend the Advance HE Aurora Leadership Development Programme for Women to understand the barriers facing women in Higher Education. TU Dublin also assigns a mentor to all leadership development programme participants.

2.2.2 The Department of Education & Skills Gender Task Force—Accelerating Gender Equality in Irish Higher Education Institutions—Gender Action Plan 2018–2020

Following publication of the HEA National Review of Gender Equality (Higher Education Authority, 2016), the Department of Education and Skills (DES, 2018) identified the review as an important first step in highlighting gender inequality in our HEIs. However, in monitoring the implementation of the HEA 2016 recommendations (Higher Education Authority, 2016), concerns were raised that progress remained exceptionally slow.

> Data trends in the universities over the period 2013-2017, show that there has been a consistently low rate of change year on year, only 1-2% each year at professor level, from a starting position of 18% female professors in 2013 to just 24% in 2017. The HEA Higher Education Institutional Staff by Gender report (2018) highlights that in 2017 only 24% of professor posts were filled by women as compared to 51% female lecturers, the entry level for academic posts in the university sector. While the number of female professors is just one metric, it is a key metric used internationally to compare countries, and it clearly highlights the extent of the problem in academia.

> Minister for Higher Education Mary Mitchell O'Connor, Gender Action Plan 2018–2020, p.2 (Higher Education Authority, 2016).

Drawing on the work of the HEA Expert Group, a 'Gender Task Force' was established to help embed the recommendations. Specifically addressing HEA recommendation 1.21—To ensure a roadmap for attainment of gender equality is developed, each institution will implement a gender action plan, the Gender Task Force outlined a Gender Action Plan 2018–2020 that identified goals, actions and targets to be integrated into the institution's strategic plan.

Review Process

The Gender Equality Taskforce formally began their work in November 2017 following the development of their Terms of Reference. The approach taken by the Gender Equality Taskforce involved the following stages:

– Review of recruitment and promotion policies and practices in HEIs conducted by an external third party, Advance HE (please see below for further information).

- Consultation with stakeholders involving face-to-face meetings with the Gender Equality Taskforce, a stakeholder consultation event and analysis of the outcomes of the consultation process.
- Data analysis of academic staff recruitment and promotion by gender and identification of gaps in data collection, particularly in regards professional, management and support staff data.
- Progress reports from HEIs on the HEA Expert Group Report recommendations and analysis of progress made at sectoral level.
- Literature review of national and international practice including challenges and emerging solutions since 2016.

Development of the three-year Gender Action Plan 2018–2020

The Gender Action Plan 2018–2020 (DES, 2018) identified specific actions and targets to accelerate progress in achieving gender equality in Irish HEIs:

1. Driving sustainable change in the form of a Centre of Excellence for Gender Equality:

   - Providing centralised support for HEIs.
   - Disseminate and share good practice and funding for innovative organisational and cultural change initiatives nationally through the HEA Gender Equality Enhancement Fund.

2. Institutional Gender Action Plans:

   - All HEIs shall submit their institutional gender action plan to the HEA and provide annual progress updates.
   - To accelerate gender balance, all HEIs shall set ambitious short-, medium- and long-term targets (1, 3 and 5 years) for the proportion of people of each gender which it aims to have at senior levels of academic and professional, management and support staff across the institution.
   - Each HEI will be held accountable for achieving their targets and performance will be incentivised through additional funding or funding consequences as appropriate, to ensure progress is constant and considerable.

3. Incentivised progress through funding mechanisms:

   - Government funding for Higher Education shall be linked to an institution's performance in addressing gender inequality.
   - To drive performance there shall be rewards for progress through funding incentives and consequences for lack of engagement or effort.

4. Recognising and embedding Gender Equality through the Athena SWAN Audit & Award:

   - HEIs shall apply for an Athena SWAN Institutional Bronze award by 2019.
   - Research Funding Institutions will require HEIs to have achieved an Athena SWAN Bronze award to be eligible to access funding.

5. Gender-proofing recruitment and promotion procedures and practices:

    – All HEIs shall ensure that there are concrete actions in their institutional gender action plans, elaborated where appropriate at discipline and business unit level, to bring their existing policies in line with good recruitment and selection practice.

6. Positive Action Interventions:

    – All HEIs shall strive for gender balance in the final pool of candidates for all competitions.
    – Each HEI is required to implement the flexible cascade model as a minimum (not a maximum) for both promotion and recruitment of academic staff and senior grades of professional, management and support staff.
    – Each HEI is required to set ambitious short-, medium- and long-term targets (1, 3 and 5 years), over and above the flexible cascade model for both promotion and recruitment of academic staff and senior grades of professional, management and support staff.
    – To enable HEIs to better monitor their progress and monitor patterns, data disaggregated by gender, contract type and broad discipline area or business unit should be collected on the number of applications, recruitments and promotions for all academic grades. This data should be submitted to the HEA annually for analysis at the sectoral and national level.
    – New and additional gender-specific posts, at appropriate levels, as well as other positive action measures, should be considered where they would be a proportionate and effective means to achieve rapid and sustainable change.
    – All HEIs shall ensure that there are concrete actions in their institutional gender action plan to address stereotyping of 'female' and 'male' roles.

The next section considers a specific example of positive action to address gender equality in senior leadership in the Irish Higher Education system, and was introduced by Minister Mary Mitchell O'Connor, who provided the foreword as part of the W-STEM project.

2.2.3   Senior Academic Leadership Initiative (SALI)

SALI is an innovative and transformational positive action initiative implemented in 2019 across Higher Education alongside the Gender Action Task Force and championed by the Minister for Higher Education. The Senior Academic Leadership Initiative aims to achieve equality of outcome in the Higher Education sector. New and additional senior academic leadership posts have been funded in areas where: there is clear evidence of significant gender under-representation; where this appointment will have significant impact within the faculty/department/functional unit and the HEI; where they would be a proportionate and effective means to achieve accelerated and sustainable change within an institution. A total of 45 senior academic leadership initiative posts are being provided across sectors (e.g. university and institute of technology) over three cycles of awards. These posts are new and additional to the sector, i.e., they are in addition to the existing Employment Control Framework (ECF), and they are funded through new and additional funding provided specifically to help address the significant gender imbalance among academic staff at senior levels.

2.3   Advance Higher Education (HE)

Advance Higher Education (HE) is a member-led education sector charity, based in Ireland and the United Kingdom, that provides support to Higher Education institutions across the world. The main purpose of this organisation is to improve Higher Education for staff, students and society by addressing systematic inequality in the HE sector (Advance, 2021a, b). Advance HE stresses the importance of enhancing teaching and learning, effective governance and leadership development as key to tackling inequality. Thus, through its equality, diversity and inclusion (EDI) work, Advance HE provides support for Higher Education institutions that include charters, professional development programmes and events, fellowships, awards, student surveys, providing strategic change and consultancy services and services through membership (Advance, 2021a, b).

One of the most important services provided by Advance HE is the development of its Equality Charters which are frameworks used across the globe to support and transform equality within Higher Education (HE) and research. Charters include the Athena Swan Charter that focuses on gender equality in HE, Race Equality Charter that focuses on representation, progression and success of minority ethnic staff and students, and the International Charter that focuses on establishing equality frameworks like Athena Swan internationally that are globally comparable but also locally contextualised (Advance, 2021c) These Charters allow organisations to apply for an award recognising their commitment to, and progress on, equality and diversity. In Ireland, awards such as Athena Swan have been noted as essential for the success of European and international funding as it shows commitment to the delivery of UN Sustainable Development Goals of quality education and gender equality (Advance, 2021b). From these charters, Advance HE provides programmes and resources to assist organisations in achieving these awards by providing guidance on gender auditing an organisation, resources such as sample surveys, as well as templates and advice for leadership and development programmes.

With its successful programmes and awards, membership of Advance HE has become global, with over 380 members across the world (Table 2.2), including members in South America and Ireland (Advance, 2021b).

In Ireland, every Higher Education Institute (Institutes of Technology, Technological Universities, Universities) are international members and have relied on Advance

Table 2.2 Advance HE
Membership 2020–21

Advance HE members 2020–21	Count of organisation
Beneficiary member	73
International global member	3
International member (inc ROI)	80
Research institute member - UK	28
UK affiliate member	112
UK member	155
Grand total	451

Source https://www.advance-he.ac.uk/advance-he-members-202
0-21

HE for resources and support in developing their own framework for development of an equality and inclusion strategy. Technology University Dublin (TU Dublin) are international members of Advance HE and are actively engaged in undertaking awards such as Athena SWAN, thus we are given access to their professional development resources, assistance in developing academic leadership programmes for all interested staff and can sponsor staff for the Aurora leadership program. By providing a gender equality charter for Ireland, Advance HE was invaluable to TU Dublin in guiding our own strategy for equality, diversity and inclusion. Advance HE provided the structure of the Athena SWAN awards and the methodology to conduct a gender audit. It also provided access to additional resources and to networks through the Aurora Leadership Programme for Women in Higher Education and the development of our own leadership programme.

2.3.1   Athena SWAN and Athena SWAN Ireland

The Athena SWAN Charter provides an accreditation framework for auditing, supporting and transforming gender equality in Higher Education Institutions. Established in 2005, the focus of the original Athena SWAN Charter was to encourage and acknowledge commitment by HE institutions to the advancement of women's careers in STEMM. Over time, the charter has evolved to go beyond STEMM areas, to include all areas of women's careers in HE such as academic and professional services roles and by focusing on representation, progression of students into academia, career journey and working environment (Advance, 2021d). At its core the Athena SWAN Charter has a series of goals:

– Helps institutions achieve their gender equality objectives.
– Assists institutions to meet equality legislation requirements, as well as the requirements and expectations of some funders and research councils.
– Uses a targeted self-assessment framework to support applicants identify areas for positive action as well as recognise and share good practice.

– Supports the promotion of inclusive working practices that can increase the retention of valued academics and professional and support staff, demonstrating an institution's commitment to an equitable working environment.

In order to provide a clear pathway and recognition of success, the Athena SWAN awards were created. By using the Athena SWAN award process, HE institutions use a targeted self-assessment framework to identify gender equality goals, areas of needed action and good practice, and provide actions to align with equality legislation as well as funding and research council requirements and promote inclusive work practices (Advance, 2021e) Different application routes apply for different levels of award (Bronze, Silver and Gold), depending on scale and size (e.g., Institution, department, small or specialised institute) and if located within/outside the United Kingdom.

The Bronze award is mainly about raising awareness and acknowledging the need for change in an organisation. The award affirms that a university or HEI is aware of gender inequality issues, has identified challenges, and has an action plan for these issues. A series of webinars about Athena SWAN are provided on the Advance HE website (https://www.advance-he.ac.uk/equality-charters/athena-swan-charter) that outlines what Athena SWAN is, how to get started, dealing with data, and undertaking gender analysis and a downloadable information pack is available on their website. Advance HE also provides information sessions and a detailed FAQs section that provides guidance to institutions on how to achieve an Athena SWAN Award. In June 2021, the Athena SWAN UK Charter has undergone further transformation to provide a new framework for new and existing members. This new charter supports greater inclusion for people in all roles, of all genders and intersectional identities (Advance, 2021e), provides greater clarification and expectations for each award level, more streamlined application, increased clarity and professionalisation of panel roles, and increased guidance and support for applicants.

In Ireland, the Higher Education Authority (HEA), mandated that all higher-level institutes in Ireland (Higher Education Authority, 2019) must apply for Athena SWAN Bronze award by end of 2019, reapply within 18 months of a failed Bronze award application, and attain Bronze award within 4 years of the first failed application. In order to achieve a Bronze award, a HEI must undertake an institutional gender self-assessment and propose a 3-year action plan based on the findings of this self-assessment. For example, for TU Dublin to be successful, it would be strongly dependent on the strength of the self-assessment which involved the support of the organisation, access to data and staff, quality of data gathered and clear interpretation of data that subsequently helped guide the future action plan. Strong support from senior levels of the organisation and beyond are essential. As noted, HEA (Higher Education Authority, 2019) issued a mandate that all institutes had to apply for Athena SWAN award, thus it was seen as an essential programme to undertake. Strong institutional support and senior management buy-in are essential for success as it signals to staff that gender equality is a strategic priority. Lack of said support and buy-in can create significant difficulties and result in failure, thus TU Dublin approached different colleges/departments/faculty within the university to set up 'Athena SWAN Champions', senior-level individuals who would represent each department and discuss and advocate for Athena SWAN to the most senior levels of the organisation.

Unsuccessful Bronze award applications do not just result from lack of support but also from lack of qualitative data, lack of analysis and interpretation of data collected, not answering all questions or missing information, not using staff consultation to inform analysis and actions, and action plans not being SMART or data informed. Thus, it is important to create a self-assessment team (SAT) that can fulfil the criteria needed for the Bronze award. Members should include a mixture of individuals with different skillsets that can help at all stages of the process. Individuals should be interested in equality and change, with a proper representation of staff included (e.g., gender, ethnicity, mobility, age, etc.).

TU Dublin recruited SAT members by asking for expressions of interest which helped gather a diverse team of skilled individuals from all levels and areas in the institute. It is important to have a mixture of academic and professional, management, and support staff to help understand the formal and informal organisational structure and culture of the institute. In TU Dublin, a formal structure for SAT was created, with working groups for each area of the application. Working groups included:

– Working Group 1 Institutional Data Analysis: focused on gathering data from the organisation for example, number and percentage of men vs. women in lecturing roles for past 5 years; number/ percentage men vs women who are junior members of staff or recently promoted.
– Working Group 2 Staff Consultation: focused on gathering quantitative and qualitative data from staff via a survey and focus groups.
– Working Group 3 Implementing Actions: considered data findings from Group 1 and 2 for possible actions for application
– Working Group 4 Communications and Outreach: promoting and raising awareness of the application, setting up a detailed website, repeatedly emailing staff and line managers to request staff to complete the survey or part take in focus groups, etc. Organising with Advance HE for seminars and workshops.

Members of each working group reported to a Group Chair, who then communicated to the overall project manager of the application. Efficient two-way communication is essential within working groups and in the overall team to identify issues and actions. Each working group had regular meetings, but also regular larger meetings were held with all working groups so progress could be reported, and future steps discussed in a transparent manner with all SAT members. Thus, effective project management for the self-assessment and action plan is essential, with clear guidance to each working group on their purpose, goals and how their contribution feeds into the application as each working group has an important part in the process.

For Working Group 1, it is important to have staff who can access institutional data such as staff from Human Resources, ICT, and/or senior management such as Office of the President (or equivalent). Data gathered in this group provides objective information about the current status of the organisation in terms of staff statistics and comparison of the national average. In the case of TU Dublin, it helped identify several issues which were then examined in the staff survey and focus groups that consequently provided SMART actions for the future action plan.

For Working Group 2, it is important to include staff who are skilled in quantitative and qualitative analysis, preferably staff that also has expertise in gender research. Surveys should include validated/previously used Athena SWAN questions with Likert scales where possible. In TU Dublin, scales, such as Career Barriers Inventory (Higher Education Authority, 2019) were adapted for use. The TU Dublin Staff Survey and focus groups focused on attitudes towards a range of areas including gender inequality, career barriers (Swanson et al., 1996), promotion, training and performance reviews, work–life balance, stress at work, work environment, etc. Samples of surveys are available online which can be used as a template that can be tailored to your organisation. Contextually specific issues can also be included such as progression to Lecturer role, which was flagged as an issue by Working Group 1, which provided interesting findings and a specific action for the TU Dublin application. Data should be analysed using statistical programmes, like SPSS or R, by someone with statistical skill to examine and interpret whether statistically significant gender differences occurred in the data. Similarly for qualitative data, an individual who is skilled in facilitating focus groups and/or interviews and analysing data is important for complementary interpretation of the data. In TU Dublin, we had several individuals in Working Group 2 involved in survey design, collection and analysis, and involved in focus group design, facilitation and analysis of results. Data collection and dissemination of results to staff by Working Group 4 was fundamental to the success of Working Group 2, with Working Group 4 maintaining close contact with Athena SWAN champions and management in order to encourage staff participation.

Thus, the campaign by Working Group 4 and analysis by Working Group 2 using a mixed-method research approach provided a rich source of information for Working Group 3 to consider future actions. For Working Group 3, members with strategic focus are essential such as those in Human Resources, senior positions, etc., as well as those with expertise in gender and equality policies. For Working Group 4, a strong team with communication and marketing skills is essential. Members in TU Dublin were from Public Affairs unit and others including student academic affairs. As noted above, each working group is not a silo, and it is important that the dependency on each working group with each other is emphasised by the project manager.

TU Dublin holds a legacy award which was granted under the TU Legacy Award Process after a constitute HEI (DIT) was successful in the application of the Bronze award using the Athena SWAN Ireland Charter as a framework for our own process. The legacy award was received in recognition of TU Dublin's commitment to advancing gender equality. The university has undertaken a significant programme of work to raise gender awareness across TU Dublin and to help create equal opportunities for career advancement. Some highlights include:

- Appointing a Director of Equality, Diversity, and Inclusion (EDI).
- Establishing a gender-balanced senior leadership committee as the highest-decision-making committee in TU Dublin.
- Providing unconscious bias training for all senior leadership and all staff.
- Commissioning an external review of all HR policies and practices to ensure that diversity and inclusion are embedded in all stages of the recruitment, selection and retention process.
- Enhancing our Leadership programme offerings; This has been updated to include an equality and diversity dimension; TU Dublin sponsors 20 women each year to participate in the Aurora Leadership Development Programme for Women in Higher Education.

2.3.2   AURORA Leadership Programme and Other Leadership Initiatives

A part of TU Dublin's successful application was enhancing leadership programme offerings to include an equality and diversity dimension. The Aurora Leadership Programme for Women developed by Advance HE for women up to senior lecturer level or professional services equivalent in HE institutions aims to support women to understand and address the barriers to career progression identified in the HEA Review of Gender Equality in Higher Education (Higher Education Authority, 2016) and the Gender Task Force Action Plan (DES, 2018). Each year, TU Dublin sponsors 20 women to attend the Aurora Leadership programme to complete the following learning modules:

- Identity, Impact & Voice.
- Politics & Influence.
- Core Leadership Skills.
- Adaptive Leadership Skills.
- Action Learning Set.

Participants have the opportunity to build their professional network across the Higher Education sector with over 100 women from the sector attending the annual Aurora programme.

Each participant is also assigned a mentor for their continued personal and professional development after the formal programme is complete. The mentoring programme enables self-directed personal and professional learning, critical thinking and the transfer of valuable organisation knowledge. The HEA expert report on Gender (Higher Education Authority, 2016) and the Department of Education and Skills Gender Action Plan (DES, 2018) have identified mentoring as an important development initiative to address the under-representation of women in senior roles in Higher Education.

2.4 Gender Equality Strategies and Policies of Irish Research Funding Organisations

In recent times the strategies of the main funding organisations in Ireland have been developed to ensure the gender element is considered in the design of all funded research.

The two main funding agencies of research in Higher Education in Ireland are the Irish Research Council and Science Foundation Ireland.

2.4.1 Gender Equality Policy and the Irish Research Council

The Irish Research Council was formed in 2012, through the amalgamation of the Irish Research Council for Science, Engineering and Technology (IRCSET), established in 2000, and the Irish Research Council for the Humanities and Social Sciences (IRCHSS), established in 2002. It is an autonomous agency of the Irish Government's Department of Education and Skills, under the aegis of the Higher Education Authority (HEA). The IRC has a mandate to fund excellent research within, and between, all disciplines. Through a variety of funding schemes, including co-funding with enterprise partners, it supports the education and skills development of excellent individual early stage researchers and cultivates agile independent researchers and thinkers. It has a remit to enrich the pool of knowledge and expertise available for addressing Ireland's current and future needs, whether societal, cultural or economic, and to deliver for citizens through collaboration and enabling knowledge exchange with Government departments and agencies, enterprise and civic society. It also provides policy advice on postgraduate education, and on more general research matters, to the HEA and other national and international bodies. It gives particular attention to the Arts, Humanities and Social Sciences.

Meanwhile Science Foundation Ireland (SFI), established in 2000, is the Irish national foundation for funding STEM research and talent within Higher Education institutions and SFI Research Centres. It aims to improve Ireland's economic competitiveness, create employment, and enhance vital areas of society, including healthcare, environment, agriculture and education. Also of relevance is SFI's remit to promote and support STEM education and engagement and create public awareness and understanding of the value of STEM to society and to the growth of the economy.

Consequently, to ensure the gender element is considered within STEM research, Higher Education, and public engagement in Ireland, it is important that each organisation includes gender equality in its high-level strategic plans, and consequent action plans.

To progress this the IRC developed a Gender Strategy and Action Plan 2013—2020 (Irish Research Council, 2013). This strategy sought to address two main issues. Firstly, the under-representation of women in STEM leading to underutilising a significant population of highly talented researchers who could be vital assets in maximising collective research intelligence and optimising creativity and innovation potential. Secondly, the need to recognise the gender dimension in the definition of research projects including those which sex and/or gender may not be relevant in terms of the research content but may have poorer results and missed opportunities if there is a failure to integrate sex and gender analysis into the design, implementation, evaluation and dissemination of the research. Informed and supported by the EU policies on gender equality in research, and by ongoing initiatives between Irish organisations and Higher Education institutions and EU partners, the IRC's strategy and action plan centred around two key objectives:

– Supporting Gender Equality in Researcher Careers.
– Objective: The Council will encourage and implement initiatives which promote equality between women and men at all stages of the researcher's career.
– Integration of sex/gender analysis in research content.
– Objective: The Council will ensure that researchers have fully considered whether their research contains a sex and/or gender dimension and, if so, that they have fully integrated it into the research content.

Additionally, the IRC committed to a process of Internal Gender Proofing such that all efforts towards the two main objectives would be a shared, overall task for the organisation as a whole. This included internal training, review of funding policies and award assessment criteria and procedures, sex-disaggregated statistical data and analysis of funding trends.

Details of the specific actions under each objective can be found in the strategy document (Irish Research Council, 2013), and a progress update from 2016 is available (Irish Research Council, 2016). The IRC provided a short summary and update of its key actions in 2018, and this newly made reference to the Athena SWAN process, as well as a new collaborative fund with the EU, through the GENDER-NET PLUS programme, which specifically called for projects relating to the interactions and interdependencies between UN SDG 5 Gender Equality and other SDGs, specifically SDG 3 Good health and well-being; SDG 9 Industry, innovation and infrastructure; and SDG 13 Climate action (Irish Research Council, 2018).

The IRC has continued its commitment to gender equality, with inclusion of a key objective to 'Build on the IRC's leading track record in promoting gender equality and the integration of the gender dimension in research, in step with evolving international best practice' under a Strategic Goal to 'Enable excellence in people, skills and ideas across all disciplines for discovery and enterprise research' within its Strategic Plan 2020–2024 (Irish Research Council, 2020a). Further information can be found in the corresponding Implementation plan (Irish Research Council, 2020b).

2.4.2   Gender Equality Policy and Science Foundation Ireland

Meanwhile, in 2016, following the publication and recommendations laid out in the HEA National Review of Gender Equality in Irish Higher Education Institutions (Higher Education Authority, 2016), Science Foundation Ireland outlined its Gender Strategy 2016–2020 (Science Foundation Ireland, 2016). The strategy was built around three strands:

– Strand 1 focused on gender equality across Science Foundation Ireland education and public engagement initiatives, with the aim of increasing the participation and interest of girls in STEM-related activities.
– Strand 2 targeted female representation within the Science Foundation Ireland funded portfolio and Science Foundation Ireland review panels. Concrete measures to achieve these targets were outlined.
– Strand 3 aimed to ensure that gender perspectives are integrated into the research content of Science Foundation Ireland-funded research programmes.

Strands 2 and 3 resonate with the key objectives of the Irish Research Council. Of particular interest to readers of this book, Strand 1 was developed based on the findings of an SFI commissioned 2014 study into the career choices of young people in Ireland (Futures, 2014). (Note, the publication appendix includes the survey questionnaire used). The study examined the key influencing factors in course selection for a representative sample of first-year undergraduates and found that 'fitting in' and being able to identify themselves in a future role was the most important factor. Career opportunities and earning potential were also identified as important, but secondary. Information about a particular course or career would not even be sought by young people if they have no affinity with the associated stereotypes, which reinforced the importance of breaking perceived stereotypes. Consequently, Strand 1 sought to increase the participation and interest of girls in STEM-related activities, and thereby their confidence in the relevance of studying STEM subjects. Actions under this strand included developing partnerships with groups that support girls to explore STEM skills informally; ensuring gender parity is addressed in the role models profiled; and delivering training to avoid unconscious attitudes or statements that may impact on girls' aspirations. Additionally, the 'SFI Discover' programme for public engagement in STEM would fund projects that aimed to increase the number of women pursuing STEM subjects and SFI would provide training in unconscious bias to SFI Discover supported education and public engagement projects. SFI also committed to ensuring its public engagement materials, activities and online content represented gender parity and challenged unconscious bias. The full list of actions relating to all three Strands are available (Irish Research Council, 2020b). These actions also resonate with the findings of another SFI study commissioned in 2015, 'Science in Ireland Barometer An analysis of the Irish public's perceptions and awareness of STEM in society' (Futures, 2014). (Note, the publication appendix includes the research methodology used). This further explored Irish awareness and trust in STEM, interest in STEM careers, as well as understanding the disengaged and disenfranchised members of the population. This study also found that the latter were more likely to be female.

The most recent SFI Strategy to 2025 (Science Foundation Ireland, 2015) focuses on three key areas, namely Excellent Research; Top Talent; Tangible Benefits. Within the 'Top Talent' area, there is an ongoing commitment to gender equality, captured within the expanded Equality, Diversity and Inclusion umbrella. Gender remains as a Key Performance Indicator, with targets of 35–40% females in STEM leadership and decision making, in keeping with Irish Government targets for gender representation for state organisations and agencies, including the Irish Research Council and the Higher Education Authority (Science Foundation Ireland, 2021).

2.5 Conclusion

Although, like elsewhere, gender equality is still a feature of the Irish Higher Education System, there have been numerous and significant policies and strategies aiming to balance gender participation in all levels and areas of the system over the last decade. In a top-down approach, commencing with EU policy, there has been a steady and perceptible increase in focus on gender equality in Ireland. Considerable leadership from the Irish Higher Education Authority has ensured that senior leaders in all Higher Education Institutions must consider gender equality as part of institutional strategic planning. The framework provided by Athena SWAN has allowed all Institutions to carry out a systematic gender audit, and to develop an action plan to address identified gender equality issues. The linking of research funding to the requirement for Athena SWAN awards has been particularly impactful.

There may often be a bottom-up approach to gender equality, with interested individuals carrying out impactful initiatives, including those involved in the W-STEM project. However, without a policy framework to address gender equality, and without meaningful engagement of senior leaders with these policies, it will not be possible to address systematic gender inequality within Higher Education in a sustainable manner—in Europe, Ireland, Latin America, or elsewhere.

Acknowledgements This work is partially supported by the European Project W-STEM, 'Building the future of Latin America: Engaging women into STEM' (Reference number 598923-EPP-1-2018-1-ES-EPPKA2- CBHE-JP). The content of this publication does not reflect the official opinion of the European Union. Responsibility for the information and views expressed in the publication lies entirely with the authors.

Appendix

Table A.1 HEA Recommendations (2016), Gender Action Plan 2018–2020 recommendations and the actions taken by TU Dublin to address these recommendations

HEA & Gender Task Force Recommendations for Higher Education Institutions					Tu Dublin implementation
HEA Objective	HEA (2016) recommendation	Lead stakeholder	Time-line/KPIS	Gender equality task force : gender action plan 2018-2020	
Leadership of higher education institutes					
1.1 To foster gender balance in the leadership of HEIs	At the final selection step, in the appointment process for new presidents (or equivalent), in so far as possible the final pool of candidates will comprise an equal number of women and men. HEIs From 2016 (including competitions already under way) If it has not been possible to achieve gender balance at the final selection step, the interview panel will account to the governing authority or equivalent for why this was not possible	HEIs	From 2016 (including competitions already under way)	To accelerate gender balance, all HEIs shall set ambitious short, medium and long-term targets (1, 3 and 5 years) for the proportion of people of each gender which it aims to have at senior levels of academic and professional, management and support staff across the institution All HEIs shall strive for gender balance in the final pool of candidates for all competitions Each HEI is required to implement the flexible cascade model as a minimum (not a maximum) for both promotion and recruitment of academic staff and senior grades of professional, management and support staff Each HEI is required to set ambitious short, medium and long-term targets (1, 3 and 5 years), over and above the flexible cascade model for both promotion and recruitment of academic staff and senior grades of professional, management and support staff	Each HEI is required to implement the flexible cascade model as a minimum (not a maximum) for both promotion and recruitment of academic staff and senior grades of professional, management and support staff. P13 'New data analysis conducted by the Gender Equality Taskforce on the estimated rate of change for the future suggests that the adoption of a flexible cascade model approach alone (as recommended by the HEA Expert Group) could take more than 20 years to achieve gender balance at professor level (i.e. minimum 40% female professors). If the vision that 'by 2026 Ireland will be a world leading country for gender equality in higher education' is to be realised, we need to look at the problem of gender diversity in academia in a new way, transformational positive action measures need to be implemented', p.2 The HEA Expert Group recommended the HEIs adopt a flexible cascade model, whereby the proportion of men and women to be recruited or promoted to a certain level is based on the proportion of each at the career level directly below. While there are some exceptions (e.g. senior lecturers in the university sector and senior lecturer 2 in the IcT sector), HEIs are broadly adhering to this model for academic staff. However, none of the HEIs have significantly exceeded the flexible cascade model threshold. Analysis shows that achieving gender balance at senior level in the university sector could take 20 years if the flexible cascade model approach alone is implemented. The introduction of a later retirement age could also slow down staff turnover. The Gender Equality Taskforce therefore recommends HEIs should ensure the flexible cascade model is used as a minimum, not as a maximum, and set ambitious short, medium and long-term targets (1, 3 and 5 years), over and above the flexible cascade model, for both promotion and recruitment of academic staff, and senior grades of professional, management and support staff. P.8

(continued)

(continued)

Leading cultural change				
1.2 To ensure HEI leaders foster a culture of gender equality in their HEI	In the appointment process for a new president, a requirement of appointment will be demonstrable experience of leadership in advancing gender equality	HEIs	Effective immediately	TU Dublin Recruitment, Selection & Appointment Policy
1.3	In the appointment process for a new vice-president, a requirement of appointment will be demonstrable experience of leadership in advancing gender equality	HEI Presidents	Effective immediately	TU Dublin Job Description Essential Criteria 'Demonstrable experience of leadership in advancing gender equality'
1.4 To lead cultural and organisational change in their area of responsibility	The deans and heads of school/ department, divisional directors and section/unit managers will be responsible for integrating gender equality in all processes and decisions made Evidence of leadership in advancing gender equality will be taken into account in appointments to these management positions	HEIs	Effective immediately	TU Dublin Job Description Essential Criteria 'Demonstrable experience of leadership in advancing gender equality'

(continued)

(continued)

1.5 To achieve gender equality in each HEI	Each HEI will, through a publicly advertised competitive process, appoint a vice-president for equality who will be a full academic member of the executive management team and who will report directly to the president	HEIs	From 2017	TU Dublin Directorate Equality Diversity & Inclusion Prof. Yvonne Galligan, VP for Equality, Diversity & Inclusion
1.6 To ensure gender balance on all key decision-making bodies	Key decision-making bodies (concerned with resource allocation, appointments and promotions) in HEIs will consist of at least 40% women and at least 40% men	HEIs	From 2016	From 2016 TU Dublin require gender balance on all university decision-making committees
1.7 To ensure gender balance on all key decision-making bodies	At least 40% of the chairs of key decision-making bodies (concerned with resource-allocation, appointments and promotions) across the HEI will be of each gender in any given year. It is expected that over a three-year period the ratio would be 50:50 women and men chairs	HEIs	By 2018	Implemented by TU Dublin 2017

(continued)

(continued)

Objective	Action		Responsible	Timeline	Recommendation	Example
1.8 To provide strategic oversight of organisational processes and policies in relation to gender equality	A gender equality sub-committee of the governing authority/body should be established	The minutes of the sub-committee will be published within the HEI	HEIs	By 2017		TU Dublin Equality, Diversity and Inclusion Committee
Embedding gender equality in organisation culture						
1.9 To support the mainstreaming of gender equality across the HEIs	Each HEI will establish an independent, academically-led gender equality forum, chaired by the vice-president for equality and comprising staff members drawn from across the HEI with sufficient influence and motivation to effect change		HEIs	By 2017	A new 'Gender Equality Enhancement Funding Call' should be set up to support innovative organisational and cultural change initiatives nationally A Centre of Excellence for Gender Equality, with a dedicated resource, shall be established by the Department under the auspice of the HEA The Centre of Excellence for Gender Equality will support the sharing of good practice by HEIs across the sector The Centre shall ensure sustainable acceleration towards gender equality in HEIs by: foster HEI collaboration and disseminate good practice; provide centralised support for HEIs; report regularly to the Minister in relation to performance of the system New Gender Equality Taskforce Objective to ensure sustainable acceleration towards gender equality in the institutions	TU Dublin Athena SWAN Self Assessment Team Athena SWAN Charter

(continued)

(continued)

1.10 To enhance the provision of support for staff members with caring responsibilities	Each HEI will establish a cross institutional working group to develop a funded structure of family leave (inclusive of maternity, paternity, parental, adoptive, and carer's leave) and develop mandatory guidelines to underpin this	HEIs	By 2017	Athena SWAN Ireland National Committee & Practitioner Network
1.11 To increase gender awareness among staff	The HEI will adopt measures aimed at actively developing gender awareness among all staff	HEIs	By 2016	TU Dublin Women Leaders in Higher Education Network TU Dublin Mentoring Programme Aurora Leadership Development Programme for Women TU Dublin Leadership Development Programme
1.12 To embed the gender dimension in teaching and learning and quality review processes	The gender dimension will be fully integrated into undergraduate and postgraduate curricula HEI Ongoing Face-to-face, unconscious-bias training will be fully integrated into initial teacher education At department level, self-assessment (departmental reviews) will include consideration of the gender dimension HEIs will include consideration of the gender dimension in the institutional quality assurance report	HEI	Ongoing	Gender equality dimension has been included in the new School review process of the TU Dublin Quality Framework TU Dublin Quality Framework The Institutional Quality Assurance Report is determined by Quality and Qualifications Ireland (QQI), there is a template which must be completed by all HEIs and this includes Equality, Diversity & Inclusion (EDI) reflections and considerations which include a measure of gender balance QQI Quality Assurance Report

(continued)

(continued)

					Researcher Career Development Framework
1.13 To embed the gender dimension in research content	Ensure that the gender dimension is integrated into all research content and provide training and support for research staff on how to do this Schiebinger, L. et al. (eds.) (2011–2015) Gendered Innovations in science, health and medicine, engineering, and environment	HEI	Ongoing		Researcher Career Development Framework
1.14 To ensure transparent distribution of work	Ensure HEI workload allocation models are transparent and monitored for gender bias on an annual basis HEIs From 2016 Evidence of this will be taken into account in the performance development reviews of managers/ supervisors responsible for setting staff workloads	HEIs	2016		TU Dublin Strategic Priority 'Establish a workload model by 2021 that enables staff to easily engage in the full breath of the University's activity'
1.15 To enable gender disaggregated data-driven decision making	A comprehensive gender disaggregated data collection system will be in place in every HEI	HEIs	2016	To enable HEIs to better monitor their progress and monitor patterns, data disaggregated by gender, contract type and broad discipline area or business unit should be collected on the number of applications, recruitments and promotions for all academic grades This data should be submitted to the HEA annually for analysis at the sectoral and national level. Similar data should also be included for senior professional, management and support staff	Annual Gender Equality Data submitted to the HEA Higher Education Data by Gender 2020

(continued)

(continued)

Recruitment & promotion practices					
1.16 To gender-proof recruitment, selection and promotion procedures and practices	The recruitment, selection, and promotion procedures currently used, will be reviewed to ensure that they are gender-sensitive	HEI	2016	A report on the recruitment, selection and promotion procedures and practices must be submitted to the EDI sub-committee of the governing authority at least once annually, and should include statistical analysis of applications, recruitment, and progression for all academic and professional, management and support staff. Relevant HEA Expert Group Objective :- To gender-proof recruitment, selection and promotion procedures and practices	HR completed an external review of recruitment and selection policy and practices to inform the TU Dublin TU Dublin HR Policies
1.17 To drive change through the use of positive action interventions for academic staff	Each HEI will introduce mandatory quotas for academic promotion, based on the flexible cascade model where the proportion of women and men to be promoted/recruited is based on the proportion of each gender at the grade immediately below	HEIs	From 2016		

(continued)

(continued)

				HEA Senior Academic Leadership Initiative (SALI) TU Dublin Senior Academic Leadership Initiative (SALI)
1.18 To drive change at professor level through the use of positive action interventions	A minimum of 40% women and 40% men to be full professors, at the appropriate pay scale	HEIs	Achieved by 2024	
1.19 To drive change through the use of positive action interventions for non academic staff	At the final selection step in the appointment process for non academic positions where the salary scale reaches or exceeds €76,000, in so far as is possible, the final pool of candidates must comprise an equal number of women and men. If it has not been possible to achieve gender balance at the final selection step, the interview panel must account to the Governing Authority or equivalent for why this was not possible	HEIs	From 2016 (including competitions already underway)	New and additional gender-specific posts, at appropriate levels, as well as other positive action measures, should be considered where they would be a proportionate and effective means to achieve rapid and sustainable change. All HEIs shall ensure that there are concrete actions in their institutional gender action plan to address stereotyping of 'female' and 'male' roles
1.20 Combat stereotyping of 'female' and 'male' roles and horizontal segregation among non-academic staff	Overtime, achieve greater gender balance at all career levels (pay grades) within the institution	HEIs	From 2016	

(continued)

(continued)

	Requirement for a gender action plan & to conduct athena swan gender audit				
1.21 To ensure a roadmap for attainment of gender equality is developed in each institution	Each HEI will develop and implement a gender action plan218 (including goals, actions and targets), which will be integrated into the institution's strategic plan and into the HEI's compacts with the HEA	HEIs	From 2016	All HEIs shall submit their institutional gender action plan to the HEA and provide annual progress updates Funding for Higher Education Institutions will be linked to an institution's performance in addressing gender inequality through the Strategic Dialogue process and System Performance Framework	Higher Education National Gender Action Plan 2018–2020 Irish Funding Bodies to Require Athena SWAN Gender Equality
1.22 To support and recognise the embedding of gender equality across all aspects of the work of HEIs	HEIs will apply for and achieve an Athena SWAN institutional award within three years	HEIs	By 2019	HEIs shall apply for an Institutional Bronze award by 2019 EIs should retain their Bronze award until such time as they obtain a Silver award. IoTs working towards TU status will be required to show evidence to the HEA, annually through their institutional gender action plans, that they are working together to build gender equality into their merger process Once a TU has been established, it shall be required to achieve a TU bronze award within three years	Athena SWAN Charter Athena SWAN Good Practice Initiatives Example – TCD Athena SWAN Gender Audit & Action Plan TU Dublin Athena SWAN

References

World Economic Forum. (2019). Global Gender Gap Report 2020. www.weforum.org/reports/gen der-gap-2020-report-100-years-pay-equality. Retrieved September 06, 2021.

European Commission. (2021). She Figures 2021. https://op.europa.eu/s/so6Z. Retrieved September 06, 2021.

Employment Equality Act. (1998). Dublin: Stationery Office. http://www.irishstatutebook.ie/eli/1998/act/21/enacted/en/html.

Gratton, L., Kelan, E., Voight, A., Walker, L., & Wolfram, H. (2007). *Innovative Potential: Men and Women in Teams*. London Business School.

Carter, N. M., & Wagner, H. M. (2011). The bottom line: corporate performance and women's representation on boards (2004–2008). *Catalyst*.

Reinert, R. M., Weigert, F., & Winnefeld, C.H. (2016). Does female management influence firm performance? Evidence from Luxembourg banks. *Financial Markets and Portfolio Management, 30*(2), (1), 113–136.

Hunt, V., Layton, D., & Prince, S. (2015). *Diversity Matters. Mckinsey & Company, 1*(1), 15–29.

Gassmann, O. (2001). Multicultural teams: Increasing creativity and innovation by diversity. *Creativity and Innovation Management, 10*(2), 88–95.

Dai, Y., Byun, G., & Ding, F. (2019). The direct and indirect impact of gender diversity in new venture teams on innovation performance. *Entrepreneurship Theory and Practice, 43*(3), 505–528.

European Commission. (2000). Science policies in the European Union, p. vii.

European Commission. (2020a). Evaluation of the strengths and weaknesses of the Strategic Engagement for Gender Equality 2016–2019 Luxembourg: Publications Office of the European Union 2020a ISBN 978–92–76–14631–5. https://doi.org/10.2838/451205.

European Commission. (2020b). Gender Equality Strategy 2020b–2025. https://ec.europa.eu/info/sites/default/files/aid_development_cooperation_fundamental_rights/gender_equality_str ategy_factsheet_en.pdf. Retrieved September 06, 2021.

García-Holgado, A., Camacho Díaz, A., & García-Peñalvo, F. J. (2019). Engaging Women into STEM in LATIN America: W-STEM Project. In TEEM'19 Proceedings of the Seventh International Conference on Technological Ecosystems for Enhancing Multiculturality (Leon, Spain, October 16th-18th, 2019 (Ed.), Conde-González, M.Á., Rodríguez-Sedano, F.J., Fernández-Llamas, C., García-Peñalvo, F.J (pp. 232–239). ACM.

García-Peñalvo, F. J., Bello, A., Dominguez, A., & Romero Chacón, R. M. (2019). Gender balance actions, policies and strategies for STEM: Results from a world café conversation. *Education in the Knowledge Society*, 20, 31–31 – 31–15, Article 31.

Higher Education Authority. (2016). HEA National Review of Gender Equality in Irish Higher Education Institutions. Dublin: Higher Education Authority. https://hea.ie/assets/uploads/2017/06/HEA-National-Review-of-Gender-Equality-in-Irish-Higher-Education-Institutions. pdf. Retrieved September 06, 2021.

DES 2018 DES. (2018). Department of Education & Skills Gender Task Force-Accelerating Gender Equality in Irish Higher Education Institutions–Gender Action Plan 2018–2020. https://hea.ie/assets/uploads/2018/11/Gender-Equality-Taskforce-Action-Plan-2018-2020.pdf. Retrieved September 06, 2021.

Advance, H. E. (2021a). Advance Higher Education–about us. https://www.advance-he.ac.uk/abo ut-us. Retrieved Auguest 28, 2021.

Advance, H. E. (2021b). Strategy 2021b-2024. https://www.advance-he.ac.uk/sites/default/files/2021b-05/Advance%20HE%20Strategy%20%E2%80%9324.pdf. Retrieved September 06, 2021.

Advance, H. E. (2021c). International Charters. https://www.advance-he.ac.uk/equality-charters/international-charters. Retrieved September 06, 2021.

Advance 2021d Advance, H. E. (2021d). Athena Swan Charter. https://www.advance-he.ac.uk/equ ality-charters/athena-swan-charter Retrieved September 06, 2021.

Advance, H. E. (2021e). The Transformed UK Athena Swan Charter. https://www.advance-he.ac. uk/equality-charters/transformed-uk-athena-swan-charter. Retrieved September 06, 2021.

Higher Education Authority. (2019). HEA statement on Athena SWAN Charter in Ireland. https://hea.ie/assets/uploads/2019/07/HEA-Statement-on-Athena-SWAN-Charter-in-Ire land-.pdf. Retrieved September 06, 2021.

Swanson, J. L., Daniels, K. K., & Tokar, D. M. (1996). Assessing perceptions of career-related barriers: The career barriers inventory. *Journal of Career Assessment, 4*, 219–244.

Irish Research Council. (2013). Gender Strategy & Action Plan 2013–2020. Dublin: Irish Research Council. http://research.ie/aboutus/publications. Retrieved September 06, 2021.

Irish Research Council. (2016). Policies and practice to promote gender equality and the integration of gender analysis in research Progress Update June 2016. Dublin: Irish Research Council. https:// research.ie/assets/uploads/2016/06/final-_progress_report_on_gender.pdf. Retrieved September 06, 2021.

Irish Research Council. (2018). Gender Strategy and Actions. Retrieved from Dublin: Irish Research Council. https://research.ie/assets/uploads/2018/08/04108-IRC-Gender-flyer-proof03-single.pdf. Retrieved September 06, 2021.

Irish Research Council. (2020a). Strategic Plan 2020a–2024. Dublin: Irish Research Council. https://research.ie/assets/uploads/2020a/01/Irish-Research-Council-Strategic-Plan-2020a-2024. pdf. Retrieved September 06, 2021.

Irish Research Council. (2020b). Implementation Plan for Strategic Plan 2020b–2024. Dublin: Irish Research Council. https://research.ie/assets/uploads/2017/05/Irish-Research-Council-Imp lementation-Plan.pdf. Retrieved September 06, 2021.

Science Foundation Ireland. (2016). Gender Strategy 2016–2020. Dublin: Science Foundation Ireland. https://www.sfi.ie/resources/SFI-Gender-Strategy-2016-2020.pdf. Retrieved September 06, 2021.

Smart Futures. (2014). SFI STEM Research. Dublin: Amarach Research. https://www.smartfutu res.ie/wp-content/uploads/2019/01/SFI-Smart-Futures-STEM-research-Final-Report-2014-1. pdf. Retrieved September 06, 2021.

Science Foundation Ireland. (2015). Science in Ireland Barometer An analysis of the Irish public's perceptions and awareness of STEM in society. SFI-Science-in-Ireland-Barometer.pdf.

Science Foundation Ireland. (2021). Shaping Our Future. Delivering Today Preparing for Tomorrow. Science Foundation Ireland Strategy 2025. Dublin: Science Foundation Ireland. https://www.sfi. ie/strategy/shaping-our-future/. Retrieved September 06, 2021.

Chapter 3
Making and Taking Leadership in the Promotion of Gender Desegregation in STEM

Mervi Heikkinen, Sari Harmoinen, Riitta Keiski, Marja Matinmikko-Blue, and Taina Pihlajaniemi

Abstract In 2016, the United Nations (UN) Member States adopted a decision on the role of the UN Educational, Scientific and Cultural Organization (UNESCO) in encouraging girls and women to be leaders in science, technology, engineering, and mathematics (STEM) and in science, technology, engineering, arts, and mathematics (STEAM). This decision poses a special challenge for many sectors in society and posits unique opportunities for women's leadership in higher education institutions (HEIs). This chapter opens by presenting views on overcoming gender segregation in STEM. The roles of women in leadership positions in the higher education STEM research areas of a large multidisciplinary university in a Nordic country are explored. The unique paths in which four of these women have progressed in their profession, position, and promotion of equality through their diverse and multiple roles within their HEI are examined. From this collection, intertwined opportunities in assuming

M. Heikkinen (✉)
Gender Studies/Values, Ideologies and Social Contexts of Education (VISE), Faculty of Education, University of Oulu, Oulu, Finland
e-mail: mervi.heikkinen@oulu.fi

University of Oulu, Pentti Kaiteran Katu 1, 90570 Oulu, Finland

S. Harmoinen
Teachers, Teaching and Educational Communities, Faculty of Education, University of Oulu, Oulu, Finland
e-mail: sari.harmoinen@oulu.fi

R. Keiski
Environmental and Chemical Engineering Research Unit, Faculty of Technology, University of Oulu, Oulu, Finland
e-mail: riitta.keiski@oulu.fi

M. Matinmikko-Blue
Centre for Wireless Communications (CWC), Faculty of Information Technology and Electrical Engineering, University of Oulu, Oulu, Finland
e-mail: marja.matinmikko@oulu.fi

T. Pihlajaniemi
Faculty of Biochemistry and Molecular Medicine, ECM-Hypoxia Research Unit, University of Oulu, Oulu, Finland
e-mail: taina.pihlajaniemi@oulu.fi

© The Author(s) 2022

F. J. García-Peñalvo et al. (eds.), *Women in STEM in Higher Education*, Lecture Notes in Educational Technology, https://doi.org/10.1007/978-981-19-1552-9_3

leadership in the promotion of gender desegregation in STEM are identified on a micropolitical levels. The chapter concludes by elaborating institutional strategies and synergies for overcoming gender segregation in higher education STEM fields from the perspective of leadership. This chapter ends with an annexed declaration useful for local policy development and practical action.

Keywords Gender · Gender equality · Gender segregation · Leadership · STEM

3.1 Introduction

Gender equality has long been acknowledged as a key driver of social and environmental development, as well as of health and well-being. Gendered innovations (Schiebinger et al., 2020) and a gender-responsible approach to scientific knowledge production (Heikkinen et al., 2020), especially in STEM, contribute to scientific excellence and quality outcomes, enhance sustainability, make research more responsive to social needs, and promote the development of new ideas, patents, and technology (Schiebinger et al., 2020). Particularly, gender-responsible scientific approaches have become imperative in scientific knowledge production and sex and gender analysis as well as gender equality plan are required for research funding such as by the EC's *Horizon Europe* (European Commission, 2021).

The EC actively promotes gender equality in research and innovation (R&I) within the European Research Area (ERA). Its overall goal is to co-create R&I environments as gender-equal and diverse, where all talents can thrive (European Commission, 2021). Furthermore, researchers are requested to integrate the sex and gender dimensions in their research projects to improve the quality and ensure the social relevance of their produced knowledge, technologies, and innovations (European Commission, 2021). Thus, gender equality is presented as an essential condition for an innovative, competitive, and thriving European economy. Gender equality is said to create more jobs and raise productivity, which are needed for green and digital transitions and to address demographic challenges (European Commission, 2021), as well as to achieve the Sustainable Development Goals (SDGs).

Gendered career progression patterns and gender segregation in the R&I sector present challenges that highlight the need to consider the influence of organisational structural factors on gender equality. In European Research Area women occupy only 24% of leading research group positions (grade A); are prevalently under-represented in STEM fields, among others, at varying degrees among the study fields; and account for less than 10% of patent holders (Directorate-General for Research and Innovation (European Commission, 2019). Vertical gender segregation in STEM as illustrated in Fig. 3.1: Vertical gender segregation in STEM, EU-28, 2013–2016, presents the proportion (%) of men and women in a typical academic career in STEM, EU-28 countries in years 2013 and 2016 (EURAXESS, 2019).[1]

---

[1] Adopted from: https://euraxess.ec.europa.eu/worldwide/japan/status-update-gender-equality-res earch-careers-europe-she-figures-2018 (accessed 25.11.2021).

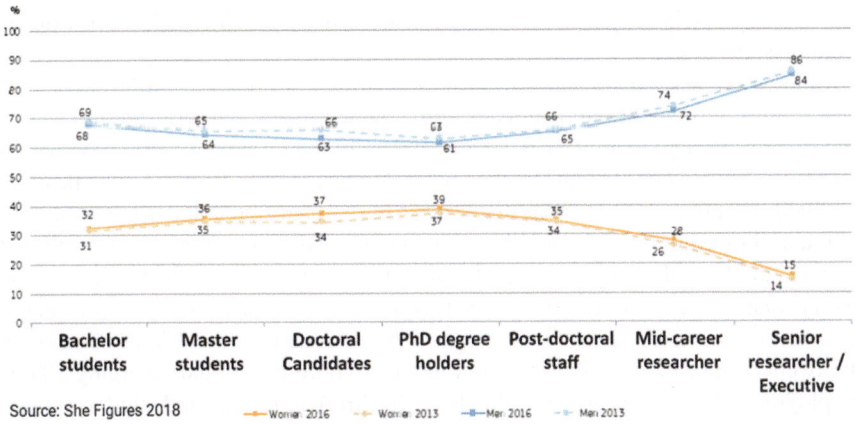

Fig. 3.1 Vertical gender segregation in STEM, EU 28, 2013 2016

The gender perspective has been discussed and scrutinised in science in various ways that range from feminist critiques of the dominant and invisible male standard in science (Haraway, 1988) to a call for increased participation and more diverse roles for women and girls in STEM (UNESCO, 2017a). Gender considerations have been found to be significant for improving the reliability of scientific knowledge production and improving individual research projects. According to Nilsen et al., (2018), science and research should be generally managed in a gender-responsible way to strengthen high-quality research and societal empowerment. They argued that gender has an impact on three elements of research: (1) the composition of the research groups; (2) the research questions in research projects; and (3) the research methods used. Thus, they suggested an approach that can be summarised as: fix the numbers, fix the institutions, and fix the knowledge - briefly, a structural change. However, women in leadership positions in academia face challenges in structures, systems, and mindsets (Alcalde & Subramaniam, 2021), that require further analysis.

3.2 Background

Over the last 20 years, there has been a visible shift in the management structures of many organisations, with more women hired in management positions, but still with little advancement in their roles (Smieszek et al., 2018). The roles and positions of women have been explored, and especially, how policy, education, international collaboration, and mentoring can support women's scientific careers and promote greater diversity among scientists. Despite the apparent significance and impact of the gender perspective on science and scientific knowledge production (Buitendijk & Maes, 2015; European Commission, 2020), it has not yet been fully integrated into research processes globally (Heidari et al., 2016), or specifically in the

Arctic (AHDRII, 2015), and not even in the European Research Area ERA (European Commission, 2021).

Studies on women's leadership have shown that women are more likely than men to present transformational leadership characteristics that motivate innovation and growth through teamwork (Cheung, 2021). In the corporate world, companies with more female board members than other companies have been found to be more profitable (Sandberg, 2019). Why, then, are so few women appointed to top positions?

Gabriele Griffin and Marja Vehviläinen concluded their extensive research that spanned the United States, the United Kingdom, and the EU (Griffin & Vehviläinen, 2021), that context matters for understanding the dynamics of persistent gender inequalities in research and innovation (R&I) throughout the Western world in the twenty-first century. Their research included more recent and emerging STEM employment fields such as biotechnology, health technology, and digital humanities, which provided useful examples of the gendered R&I career struggles that are typical under the current regime of academic capitalism. The Nordic countries Finland, Sweden, and Norway have been ranking high in both the European Gender Equality Index and the European Innovation Scoreboard. However, specific everyday experiences of women working in R&I articulate persistent gender inequalities in the academe (Griffin & Vehviläinen, 2021). Each of the four researcher career stages—doctoral student, postdoctoral researcher, researcher/university lecturer, and research director/full professor—have different gender inequalities at play. Their interplay contributes to the durable inequalities, such structural elements and practices that are hard to be recognised and articulated, but what accompany women's research careers in the Nordic countries. Griffin and Vehviläinen (2021) write:

> The Nordic version of the gender paradox in R&I (with the underrepresentation of women in top R&I positions and STEM fields) has to develop a gender equality that supposedly already exists (Martinsson et al., 2016, 1)

Women's careers are often more disrupted than men's, particularly in the new technology-driven knowledge economy. Griffin and Vehviläinen (2021) described the shift as that from a linear career path to an increasingly flexible one, which may even consist of parallel career paths. They presented the shift as having had a specific impact on women's careers, and identifying such impact makes for both a timely and urgent research topic. Those fewer female professors do provide important mentoring for younger female colleagues, but since they themselves are always in a significant minority, they can achieve only a limited elevator effect through these practices (Griffin & Vehviläinen, 2021).

3.3 Design of the Study

The recent studies both provide detailed understanding of gender segregation at the local level (Kaleva et al., 2019), and also identify and discuss on promising solutions (Ulvinen et al., 2021). This study complements the previous studies with insights on roles and paths of women leaders in STEM fields at the university. The purpose is to understand better how both gender and gender equality promotion are intertwined in their academic career paths in the STEM fields. For this purpose, four women in top leadership positions in different disciplinary areas at a Nordic higher education institution (HEI) were selected and invited to contribute to a chapter and joint elaboration of the matter. Three of them were originally educated in a STEM discipline, and one, in life science. All their work relates to STEM fields when they participated in this study. Thus, the study method forms a novel application of autobiographical study (Järvelä, 1996) consisting of publicly available career stories online, usually in a concise form. All career stories are contributing to a collective biography (Davies & Gannon, 2006) of women in STEM leadership and an organisational case study (Heikkinen, 2012) as an overall attempt to understand organisational culture and situations for structural change.

These women in top leadership positions in STEM fields were approached to ask them if they were interested to co-author this chapter and to request their permission for us to utilise various publicly available materials on their careers and on them as STEM role models for this study. All four women accepted both the invitation and use of their public materials, and two of them expressed interest in co-creating Wikipedia profiles that they did not yet have.

This chapter explores the roles and contributions of women in leadership positions in higher education STEM fields of a large multidisciplinary university in a Nordic country. Based on the career paths of these four women leaders in STEM fields, the chapter explores the individual situatedness professionally, positionally, and in relation to the promotion of equality within an HEI.

3.4 Professional Profiles of Women in Leadership Position in STEM Fields

This chapter introduces four women in leadership positions in STEM fields at the University of Oulu. They are Vice-Rector in Research Taina Pihlajaniemi, Director of Sustainability and Regulation at 6G Flagship Marja Matinmikko-Blue, Dean of the Faculty of Technology Riitta Keiski, and Education Dean of the Faculty of Education Sari Harmoinen.

The four introductions are based on already existing public profiles on the women's accomplishments, such as their profiles on the University of Oulu website and their Wikipedia profiles.[2] Additional online sources on their academic career-related activities, such as the university news and gender equality-related publications, interviews, and videos were also viewed to gain a broader understanding of their current and past leadership positions. From these sources, short descriptions of

---

[2] European Science and Innovation Commissioner Carlos Moedas discusses on the International Day of Women and Girls in Science, 11th of February 2019 how Wikipedia profile could be a one way to create more visibility for women's STEM careers. Source: https://youtu.be/YSgKLmCnJSE.

their backgrounds, career paths, research interests, and tasks related to their leadership were drawn together. Their professional public image, leadership in diverse tasks, and impacts on gender and STEM-related matters were our particular areas of interest.

3.4.1 Medical Biochemistry

Dr. Taina Annikki Pihlajaniemi is a Finnish professor of medicine and collagen researcher. She is the Vice-Rector in Research at the University of Oulu for 2010–2025. In 2012–2017 she was a director of the Academy of Finland's Center of Excellence in Cell-extracellular matrix research, which consisted of six research groups.

From 1996 to 2009, Dr. Pihlajaniemi was the scientific director of Biocenter Oulu and the director of the graduate school. Before assuming her position as Vice-Rector in Research, she served on the Board of the CSC—Center for Science Information Technology for 12 years, in the last four years of which she also chaired the organisation. Pihlajaniemi also served as the Founding Director of Biocenter Finland in 2006–2007, was a member of the Health Committee of the Academy of Finland in 1998–2003 and was a member of the Finnish Science and Technology Council chaired by the Prime Minister.

Dr. Pihlajaniemi graduated from the University of Oulu in 1981. From there, she went to Rutgers Medical School in the United States for three years as a postdoctoral researcher. In 1987, she had her firstborn. The same year, she became a Doctor of Medicine in Oulu and was appointed docent. In 1990, she received a professorship in Medical Biochemistry.

Dr. Pihlajaniemi is currently the Chairperson of the Equality and Non-Discrimination Board and has been leading a process of updating the Gender Equality and Non-Discrimination Plan of the University of Oulu for years 2022–2023.

Sources

Wikipedia: https://fi.wikipedia.org/wiki/Taina_Pihlajaniemi.

University of Oulu: https://www.oulu.fi/university/researcher/taina-pihlajaniemi.

News: https://www.oulu.fi/yliopisto/uutinen/pohjoispohjanmaan-palkinto-pihlajani emelle.

Finnish Union of University Professors, Vimeo: https://vimeo.com/32626134.

3.4.2 Telecommunications

Dr. Marja Matinmikko-Blue (until 2017, Matinmikko; born 1979 in the Rovaniemi countryside) is a Finnish researcher. She is a docent at the University of Oulu and has been the Research Coordinator in 2018–2021 and the Director of Sustainability and Regulation since October 2021 in the 6G Flagship at the University of Oulu.

Dr. Matinmikko-Blue graduated from the University of Oulu with a master's degree in Industrial Engineering and Management in 2001. She earned a licentiate in Communications Engineering in 2007, a doctorate in Communications Engineering in 2012, and a doctorate in Industrial Engineering and Management in 2018. Her research interests are interdisciplinary, combining technology, business, and regulatory perspectives in the development of ICT solutions. Since then, Matinmikko-Blue's main research topics have been related to the regulation of radio frequencies, future 6G systems, and sustainable development.

In 2013, Finnish Foundation for Technology Promotion (Tekniikan Edistämissäätiö TES) selected Dr. Matinmikko-Blue as the Young Scientist of the Year. The panel selected her for forging active and successful collaboration between research, industry, and administrations to develop the mobile networks of the future.

Dr. Matinmikko-Blue worked at Nokia as an intern in 1999 and 2000, at the University of Oulu as a teaching assistant in 2000, and at VTT Technical Research Centre of Finland Ltd. (VTT) from 2001 to 2016 as a research scientist, senior scientist, and project manager. She joined the University of Oulu's Center of Wireless Communications in 2016.

She has published more than 170 scientific publications and prepared more than 150 contributions to national, European, and international regulatory bodies; chaired cognitive radio system studies at United Nations based International Telecommunication Union Radiocommunication Sector (ITU-R). She has been a lead writer in the White Paper on 6G drivers and the UN SDGs, a contributor to the linking of mobile communications with the UN SDGs and the mapping process and the further elaboration of sustainability as a challenge and driver for novel ecosystemic 6G business scenarios (Matinmikko-Blue et al., 2020).

Sources

Wikipedia: https://fi.wikipedia.org/wiki/Marja_Matinmikko-Blue.

University of Oulu: https://www.oulu.fi/university/researcher/marja-matinmikko-blue.

News: https://www.oulu.fi/6gflagship/news/director-marja-matinmikko-blue.

Shaking up Tech: https://shakinguptech.aalto.fi/uratarinat/marja-matinmikko-blu eWomen Women In Digital: https://www.youtube.com/watch?v=neUjTsZImOg.

3.4.3 Environmental and Chemical Engineering

Dr. Riitta Liisa Keiski (born Kalliokoski) is a Finnish professor. She holds a Professorship in Mass and Heat Transfer at the University of Oulu and has been the Dean of the Faculty of Technology since 2015.

Dr. Keiski graduated from the University of Oulu with a master's degree in Process Engineering in 1981, a licentiate in Technology in 1984, and a doctorate in Process Engineering in 1991, with her research focus on heterogeneous catalysis, environmental engineering, and reactor design. Since then, her main research topics have been related to sustainable development and green chemistry and technology. She also served as a Vice-Rector of the University of Oulu in 2006–2009.

At the start of 2011, the Finnish Association of Professors elected Dr. Keiski Professor of the Year 2010 for being a researcher, developer of postgraduate education, and highlighter of issues related to research ethics and equality. Her research interests include globally important research topics related to sustainable development and green chemistry and technology. In 2013, she was awarded a research grant by the Tapani Järvinen Environmental Technology Fund.

Dr. Keiski was awarded an honorary doctorate by Åbo Akademi University in 2018. She has also received *Doctor Honoris Causa* honors from the National Engineering University in Lima, Peru in 2015 and the Corvinus University of Budapest in Hungary in 2014.

Dr. Keiski served as a president of NORDTEK in 2017–2019 and as a member of the Board in 2016–2020. She was a member of the Research Council of Natural Sciences and Engineering of the Academy of Finland from 1998 to 2000 and its Chairperson from 2001 to 2006, a Vice-Chairperson of the Research Ethics Advisory Board in 2007–2010, a member in 2010–2013, and a Chairperson from 2019 to 2022.

During the years 2010–2012 Riitta Keiski has served as a Chairperson of the Gender Equality Board of the University of Oulu. Gender equality mainstreaming has been part of her leadership and she has acted as a role model for girls and inspiration for students in green engineering and sustainability (see e.g. Saavalainen 2020). Besides, she has been a member of the working groups for Gender Equality Plans of the Academy of Finland and the University of Oulu.

Sources

Wikipedia: Riitta Keiski https://fi.wikipedia.org/wiki/Riitta_Keiski.

University of Oulu: https://www.oulu.fi/yliopisto/tutkija/riitta-keiski.

Finnish Union of University Professors, Vimeo: https://vimeo.com/32625698.

Science for Girls, interview: https://www.tiedettatytoille.fi/riitta-keiski/.

3.4.4 Mathematics and Physics

Dr. Sari Harmoinen is a lecturer on subject studies, mathematics, and physics at the Faculty of Education of the University of Oulu. She was appointed as docent on STEAM teaching and learning at the University of Turku 2020. Before teaching at the University of Oulu, Dr. Harmoinen was a high school mathematics and physics lecturer at the city of Oulu and a secondary school mathematics, physics, and chemistry. As the Education Dean of the Faculty of Education at the University of Oulu in the past six years, she has been responsible for all curriculum development and for the quality of education at the faculty. Together with her colleagues, she is developing two minor study subjects: Sustainable Development Studies (SDS) and STE(A)M. The SDS is expected to be launched in 2022.

Dr. Harmoinen has contributed widely to the quality and development of education both nationally and internationally. She was a member of the EUA peer group in 2019; Vice-Chair of FINEEC's Evaluation Council and member of FINEEC's Higher Education Evaluation Committee in 2018–2022; and a member of the Teacher Education Development Programme of the Ministry of Education and Culture.

According to the study on university students' readiness for social activity in climate actions (Harmoinen et al., 2020), students are interested in climate change but need reliable fact-based information on it. However, students are already considering contributing to climate actions.

Sources

Wikipedia: https://fi.wikipedia.org/wiki/Sari_Harmoinen.

University of Oulu: https://www.oulu.fi/yliopisto/tutkija/sari-harmoinen.

News: https://www.oulu.fi/fi/uutiset/ilmastonmuutospuhe-kyllastyttaa-nuoria-mutta-saa-aikaan-myos-tekoja.

3.5 Lessons Learned on Gender and STEM Leadership

In the previous chapter, the successful and unique careers of four professional women in STEM and related fields at the University of Oulu were briefly presented based on already existing publicly available online materials. These women have been working as researchers and teachers, research group leaders, innovators, role models, supervisors, and supporters of younger female STEM professionals. Their scientific achievements in their own academic research areas—in medical biochemistry, telecommunications, environmental and chemical engineering, and in teaching of mathematics and physics—have been remarkable.

Furthermore, they have been elected to positions at the very top of their entire university: as the Vice-Rector in Research (Taina Pihlajaniemi), the Dean of the Faculty of Technology and former Vice-Rector in Education (Riitta Keiski), Director in Sustainability and Regulation of the 6G Flagship (Marja Matinmikko-Blue), and Education Dean of the Faculty of Education (Sari Harmoinen). As part of these listed administrative academic duties, they have also put in efforts to promote gender equality in STEM fields in various projects and campaigns and even more broadly, as one of them has chaired the Gender Equality and Diversity Board (GEDB) of the university, and one is the current Chair of the GEDB.

By elaborating these four unique careers, we were able to compile these women's multiple international professional roles in STEM fields and in relation to gender equality promotion in their university. As a result of this collective action their individual leadership roles as influential women in STEM fields become even more visible. Gradually it becomes apparent, that through their professions, institutional positions, duties, and responsibilities they were influencing further on attraction, access, and retention of other women in their STEM careers, as well as sustainable development in general and UN SDG's in particular.

These women are in their 40 s, 50 s, and 60 s. They all originate from Northern Finland and from the sparsely populated countryside municipalities, from places such as Haapajärvi, Halsua, Rovaniemi, and Rovaniemi countryside. They all have extensive international experience through their research and development projects. Despite their international academic careers and networks, they are affiliated to the University of Oulu. Two of them have built their careers at the University of Oulu and two of them have built a career first outside the university—one in the public sector and the other, in a state owned non-profit company. The following table presents the main public leadership channels of the four women working in STEM and related fields, as they appear in the public data: academic professions, administrative positions, and appointments to gender equality promotion tasks and projects (Table 3.1).

Many women in STEM fields receive requests to join various projects and campaigns to attract more women into the field (Brunila et al., 2005). Such campaigns and projects as the Shaking up Tech, Women in Digital, Science for Girls, Women in Mining, Women IT in addition to W-STEM appear in data. These projects include strategies for structural transformation through implementation of systematic institutional measures, which quite blatantly aim to expose gender as an organising principle at the institution (Garcia-Holgado et al., 2020; UNESCO, 2017b; EIGE 2016). Such transformative practices include gender mainstreaming in all institutional actions, gender equality planning (GEP), as well as sex and gender analysis of research projects (European Institute for Gender Equality (EIGE, 2016). However, these tasks require at least human resources, so leadership to improve institutional gender equality is needed and in addition to that a collective, systematic commitment for its actual progression and for implementation of GEP.

Two of the leaders have chaired the Gender Equality and Diversity Board (GEDB) of the University of Oulu. The main role of the GEDB is to draft the university's GEP,

Table 3.1 Channels for women's leadership in STEM—profession, position, and promotion of gender equality

Name	Profession	Position	Promotion of gender equality
Dr. Taina Pihlajaniemi	Professor in Medical Biochemistry	Vice-Rector in Research	Chair of the Gender Equality and Diversity Board
Dr. Marja Matinmikko-Blue	University Researcher in Telecommunications	Director in Sustainability and Regulation of the 6G Flagship	Shaking up Tech, Women In Digital
Dr. Riitta Keiski	Professor in Chemical Engineering (Mass and Heat Transfer)	Dean in the Faculty of Technology, Former Vice-Rector in Education	Former Chair of the Gender Equality and Diversity Board, Science for Girls, Women in Mining
Dr. Sari Harmoinen	University Lecturer on Physics and Mathematics Didactics	Education Dean in the Faculty of Education	Women IT, W-STEM

to monitor its implementation and to revise it every other year. The GEP includes an evaluation of the current gender equality situation, statistical information on the number of human resources and salary levels in different job categories and selected institutional gender equality issues for improvement with defined measures and targets, timetable and named parties responsible for implementing those measures, and indicators to measure improvements towards targets (EIGE, 2016).

In addition to gender equality promotion women have been taking active roles in development activities related to the UN Sustainable Development Goals (UNESCO, 2017a) in universities. This has become apparent for instance in the White Paper on 6G Drivers and the UN SDGs led by Matinmikko-Blue et al., 2020), which describes the role of telecommunications in Agenda 2030. Also, minors in the Sustainable Development Studies (25 credits) and STE(A)M Studies (25 credits) under planning led by Sari Harmoinen are fostering educational capabilities for future professionals that will help them overcome their STEM subject silos and discuss global issues such as climate change (Harmoinen et al., 2020) on a transdisciplinary manner. Furthermore, supervision of postgraduate students in the Development of Sustainability Assessment Tool and Criteria for early stage process design projects (Saavalainen, 2020) has been in the core of activities in Riitta Keiski's academic work.

It is quite challenging to get women appointed to the top management positions in STEM-related fields and also more broadly to the top administration at the University of Oulu. First, it is highly likely that due to fierce competition the top positions will be filled with a candidate who is selected among the majority group of professionals from STEM fields. However, in those STEM research fields, which have the largest faculties in the University of Oulu, women are underrepresented, increasing likelihood that the minority position can create unconscious biases and thus hindrances for their career promotion, and against their access to the top positions (Frith, 2015) unless addressed by institutional support mechanisms (Husu et al., 2011).

3.6 Discussion—Strategies and Synergies for Women in Leadership

These four professionals have been and are important role models, encouragers for the STEM fields, and public speakers for campaigns for more gender equality in academia. They have built networks and alliances with SSH disciplines at the institutional, national, and international levels. Two of our focus women have a long-term affiliation with the university. In addition to an excellent academic record being familiar with one's Higher Education Institution, known and respected by colleagues may raise one's chances of having a leadership position there. Additionally a mentor's support is valuable (Husu et al., 2011). Two of the current women leaders have spent considerable time working outside the university right after their graduation providing them invaluable experience. Contacts to regional actors, municipalities, companies, and other research institutions are assets for career building also within University and for a leadership position in STEM fields.

Within the University of Oulu, institutional awareness of gender segregation as the underrepresentation of women in the STEM fields as more of a sustainability-related problem has increased. Global actions for gender desegregation are needed in both the STEM and SSH fields (Microsoft, 2017a). Identified barriers to women's entry into the STEM fields (UNESCO, 2017a) are just parts of the problem and thus, are enticing to solve for a start. Detailed understanding of the problem both on European (Bettio & Verashchagina, 2009) a national (Brunila et al., 2005; Microsoft, 2017b) and at the local level has increased (Pursiainen et al., 2018; Kaleva et al., 2019). Promising practices to solve the situated challenges have been identified and discussed further by various locally influencing actors (Brunila et al., 2005; Heikkinen et al., 2014; Ulvinen et al., 2021) during past two decades.

The W-STEM project has enabled networking, collaboration, and further articulation of the locally appearing challenges but also formulation of aspired-for solutions within an HEI. Synergies within our own local university, with W-STEM consortium partners and more extensively, through international networking have been and are empowering. Our local W-STEM work related to theoretical and conceptual development is contributing to closing the gender gap in STEM. This practical work took the form of articles published in the Arctic Yearbook 2021, entitled "Intersectional Gender-Responsibility in STEM: Co-Creating Sustainable Arctic Knowledge Production" (Heikkinen et al., 2020) and "The SAGA of Sustainable Development: Gender and Other Differences in Knowledge Production of HEIs in the STEM Field". The latter is an analytical project description published in the Finnish Gender Studies journal called Sukupuolentutkimus (Heikkinen et al., 2021). Furthermore, a piece of an academic discussion on an institutional policy enforcement aimed at structural transformation, "Gender Equality in STEM Programs: A Proposal to Analyse the Situation of a University about the Gender Gap", was published in the IEEE Global Engineering Education Conference proceedings (Garcia-Holgado et al., 2020).

3.7 Conclusion

Gender-based differences in participation in science, technology, engineering, and mathematics (STEM) fields have been apparent in higher education for decades (Brunila et al., 2005). According to the United Nations Educational, Scientific and Cultural Organization (UNESCO, 2017a), in (2014–2016), just around 30% of female college students worldwide selected STEM-related areas as their areas of study. The lowest proportion of women studied technology, manufacturing and construction, science, mathematics and statistics, and information and communication technology (ICT). One topic that has been brought up in literature is the Nordic gender paradox— a behavioural pattern among Western women of tending not to seek careers in STEM despite their success in school, due to which women remain a minority in STEM (Griffin & Vehviläinen, 2021). Within this broad global challenge, identified barriers to women's entry in STEM fields (Microsoft, 2017a) are particularly challenging to solve. Global actions are needed to address gender segregation in both the STEM and Social Sciences and Humanities (SSH) fields by solving the identified root causes of the problem (Bettio & Verashchagina, 2009).

How could we provide appropriate support to our W-STEM project partners solving issues on gender segregation in Latin America, when we have not been able to solve them appropriately in Europe, including in global gender equality leaders Nordic countries? Certainly, we can learn from each other and hopefully, from our mistakes, so that we would not repeat them, but rather co-create new gender responsible and sustainable ways of seeing, being, and becoming with STEM and STEAM (Heikkinen et al., 2020) and further making and taking leadership.

In the W-STEM project, we have discussed with the consortium partners on critical science political issues such as the appropriateness to attract more women into STEM fields as change agents, as well as related geographical, societal, and cultural issues. Due to the course of W-STEM project we attempted to generate inspiration for locally tailored, evidence-based approaches for gender desegregation both in Latin America and in Europe to support such work also elsewhere. It seems that it is the way forward—to think globally and act locally.

Globally available online gender equality resources provided by UNESCO and EU have been beneficial for the local work. The UN STEM and Gender Advancement (SAGA) tool (UNESCO, 2017b) offer remedies for addressing gender equality issues in the STEM fields within an institution, including leadership. The European Institute for Gender Equality (EIGE) provides useful applicable promising practices for institutional gender equality work through the Gender Equality in Academia and Research (GEAR) tool (EIGE, 2016). These SAGA and GEAR tools could be systematically utilised as part of the institutional GEP process, but most importantly in addition to leadership, the institutional gender equality work needs committed people, a community of practitioners—CoP's, that would benefit on interorganisational networking, support, and exchange of promising practices (Thomson et al., 2021).

Therefore, it is worth sharing that the University of Oulu launched its own Declaration on Responsible Science (Appendix) in 2020. The declaration includes gender responsibility, as follows:

> *We take into account the implementation of gender equality and diversity in the composition of research groups and research decision-making. We improve the scientific quality and the societal relevance of the produced knowledge, technology, and innovation by integrating gender analysis in research content.*

Local Declaration on Responsible Science and European Responsible Research and Innovation (RRI) approach,[3] as well as diverse ways to improve research integrity (Mejlgaard et al., 2020), pave the way for further intersectional gender equality mainstreaming at the HEIs. These approaches provide a set of useful tools for all leaders who take time from academic research for gender equality promotion at their institutions, for both women, men, and people who identify themselves differently. Thus, the administrative challenge is to provide appropriate academic merit for an individual on those institutional gender equality promotion efforts since they require serious personal input and leadership. Besides creating an equal culture and climate in HEIs is important for health and well-being of academics, it is also a profound matter in co-creating excellence in the academic outcomes.

Acknowledgements The authors acknowledge the Erasmus+ Programme of the EU in its Key Action 2, "Capacity-building in Higher Education." We also thank Project W-STEM, "Building the future of Latin America: Engaging women into STEM" (Reference number 598923-EPP-1-2018-1-ES-EPPKA2- CBHE-JP), for the collaboration and important work in the field. The authors are solely responsible for the information and views expressed in this paper.

Appendix

Declaration on Responsible Science

The University of Oulu promotes open and responsible science, responsible conduct of research, sustainable development, equality and non-discrimination in research, and responsible research assessment.

We follow the responsible conduct of research, the principles that are endorsed by the research community, that is, integrity, meticulousness, and accuracy in conducting research, and in recording, presenting, and assessing the research results.

---

[3] Responsible Research and Innovation tools: https://rri-tools.eu/ Accessed 29.11.2021.

We promote sustainable development in all our research activities and take into account its ecological, cultural, social, and economic dimensions.

We are committed to freedom of scientific research and to promoting open access to research knowledge.

We take into account the implementation of gender equality and diversity in the composition of research groups and research decision-making. We improve the scientific quality and the societal relevance of the produced knowledge, technology, and innovation by integrating gender analysis in research content.

We are committed, by signing the Finnish Declaration for Open Science and Research 2020–2025, to: promote openness as a fundamental value throughout the research community and its activities, strengthen societal knowledge base and innovation, and improve the quality of scientific and artistic research outputs and the educational resources based on them, and the fluid mobility and impact of research outputs throughout society.

We are committed, by signing the DORA Declaration, to the development of research assessment and responsible use of metrics. When assessing research, we take into account the value of all research results as well as a wide range of different indicators, including qualitative indicators such as the impact on society.

We support the goals of the European Open Science Cloud (EOSC) to bring together open science data services and data produced in the EU to make the use of services and the sharing of research results as simple and easy as possible.

References

Responsible conduct of research and procedures for handling allegations of misconduct in Finland (Finnish Advisory Board on Research Integrity).

The European Code of Conduct for Research Integrity (European Federation of Academies of Sciences and Humanities).

The United Nations Sustainable Development Goals.

Sustainability and responsibility theses of the UNIFI (Universities Finland).

Bonn Declaration on Freedom of Scientific ResearchGender Equality (Responsible Research and Innovation Tools)Promotion of equality and non-discrimination at the Academy of FinlandVademecum on Gender Equality in Horizon 2020Declaration for open science and researchDORA DeclarationEOSC Declaration.

Source Responsible research.

https://www.oulu.fi/en/science-arctic-attitude/responsible-research.

References

AHDRII, Nordic Council of Ministers. (2015). Arctic human development report: regional processes and global linkages. https://doi.org/10.6027/9789289338837-15-en.

Alcalde, M. C., & Subramaniam, M. (2021). Women in leadership in academe still face challenges in structures, systems and mind-sets (opinion). Inside Higher Ed. https://www.insidehighered.com/views/2020/07/17/women-leadership-academe-still-face-challenges-structures-systems-and-mind-sets.

Bettio, F., & Verashchagina, A. (2009). Directorate-general for employment, social affairs and equal opportunities European Commission. Gender segregation in the labour market. Root causes, implications and policy responses in the EU. European Commission's Expert Group on Gender and Employment (EGGE). https://op.europa.eu/en/publication-detail/-/publication/39e67b83-852f-4f1e-b6a0-a8fbb599b2.

Brunila, K., Heikkinen, M., & Hynninen, P. (2005). Difficult but doable. Good practices for equality work. University of Oulu, Kajaani University Consortium. Kainuun Sanomain Kirjapaino Oy, Kajaani. http://www.kajaaninyliopistokeskus.oulu.fi/proj/womenit/difficult_but_doable.pdf.

Buitendijk, S., & Maes, K. (2015). Gendered research and innovation: Integrating sex and gender analysis into the research process. Advice Paper, 18. https://www.leru.org/publications/gendered-research-and-innovation-integrating-sex-and-gender-analysis-into-the-research-process.

Cheung, F. M. (2021). The "state" of women's leadership in higher education. *International Briefs for Higher Education Leaders nbr, 9*. https://www.acenet.edu/Documents/Womens-Rep-in-Higher-Ed-Leadership-Around-the-World.pdf.

European Commission. (2021). Gender equality in research and innovation. https://ec.europa.eu/info/research-and-innovation/strategy/strategy-2020-2024/democracy-and-rights/gender-equality-research-and-innovation_en.

Davies, B., & Gannon, S. (2006). *Doing collective biography: Investigating the production of subjectivity*. Open University Press.

Directorate-General for Research and Innovation (European Commission). (2019). *She figures 2018*. Publications Office of the European Union. https://doi.org/10.2777/936.

Directorate-General for Research and Innovation (European Commission). (2020). *Gendered innovations 2: How inclusive analysis contributes to research and innovation - Policy review*. Publications Office of the European Union. https://doi.org/10.2777/316197.

EURAXESS. (2019). Status update of gender equality in research careers in Europe: She figures 2018. https://euraxess.ec.europa.eu/worldwide/japan/status-update-gender-equality-research-careers-europe-she-figures-2018

European Institute for Gender Equality (EIGE). (2016). Gender Equality in Academia and Research (GEAR) tool. https://eige.europa.eu/gender-mainstreaming/toolkits/gear

Frith, U. (2015), Understanding unconscious bias. https://royalsociety.org/topics-policy/publications/2015/unconscious-bias/

Garcia-Holgado, A., Mena, J., Garcia-Peñalvo, F. J., Pascual, J., Heikkinen, M., Harmoinen, S., Garcia-Ramos, L., Peñabaena-Niebles, R., Amores, L. (2020). Gender equality in STEM programs: A proposal to analyse the situation of a university about the gender gap. In *IEEE Global Engineering Education Conference* (pp. 1824–1830). http://hdl.handle.net/10366/142961.

Griffin, G., & Vehviläinen, M. (2021). The persistence of gender struggles in Nordic research and innovation. *Feminist Encounters: A Journal of Critical Studies in Culture and Politics, 5*(2). https://doi.org/10.20897/femenc/11165.

Haraway, D. (1988). Situated knowledges: The science question in feminism and the privilege of partial perspectives. *Feminist Studies, 14*, 575–599. https://doi.org/10.2307/3178066.

Harmoinen, S., Koivu, K., & Pääsky, L. (2020). University students' readiness for social activity in climate actions. *Discourse and Communication for Sustainable Education, 11*(1), 134–152. https://doi.org/10.2478/dcse-2020-0012.

Heidari, S., Babor, T. F., De Castro, P., Tort, S., & Curno, M. (2016). Sex and gender equity in research: Rationale for the SAGER guidelines and recommended use. *Research Integrity and Peer Review, 1*, 2. https://doi.org/10.1186/s41073-016-0007-6.

Heikkinen, M. (2012). *Sexist harassment as an issue of gender equality politics and policies at university. Academic Dissertation*. University of Oulu.

Heikkinen, M., Pääsky, L., & Harmoinen, S. (2021). Kestävän kehityksen SAGA—Sukupuoli ja muut erot korkeakoulujen STEM-alojen tieteellisen tiedon tuottamisessa. [The SAGA of Sustainable Development—Gender and other differences in a knowledge production of HEIs STEM fields]. *Sukupuolentutkimus, 2,* 38–43.

Heikkinen, M., Okkonen, E., & Selesniemi, J. (2014). Segregaation tila Pohjois-Pohjanmaalla [The stage of (gender) segregation at the Northern Ostro-Bothnia]. http://segregaationtila.blogspot.com/.

Heikkinen, M., Pihkala, S., Pääsky, L., & Harmoinen, S. (2020). Intersectional gender-responsibility in STEM: Co-creating sustainable arctic knowledge production. In: Heininen, L., Exner-Pirot, H., Barnes, J. (Eds.), Arctic Yearbook 2020. Akureyri, Iceland: Arctic portal (pp. 175–188). https://arcticyearbook.com/images/yearbook/2020/Scholarly-Papers/9_Heikkinen_et_al.pdf.

Husu, L., Hearn, J., Lämsä, A. M., & Vanhala, S. (Eds.). (2011). Women, management and leadership – Naiset ja johtajuus. NASTA Women's leadership project final report. Hanken School of Economics Research Reports 72. https://wiki.aalto.fi/pages/viewpage.action?pageId=95748875&preview=/95748875/96207129/Nasta_loppuraportti.pdf.

Järvelä, M. L. (1996). *Why hath this lady writ her own life...?: Auto/biography from feminist perspectives.Oulun yliopisto, kasvatustieteiden tiedekunta* (Vol. 2). Northern Gender Studies Publication.

Kaleva, S., Pursiainen, J., Hakola, M., Rusanen, J., & Muukkonen, H. (2019). Students' reasons for STEM choices and the relationship of mathematics choice to university admission. *International Journal of STEM Education, 6,* 43. https://doi.org/10.1186/s40594-019-0196-x.

Martinsson, L., Griffin, G., & Giritli-Nygren, K. (Eds.). (2016). *Challenging the myth of gender equality in Sweden.* Policy Press. https://doi.org/10.1332/policypress/9781447325963.001.0001.

Matinmikko-Blue, M., Aalto, S., Asghar, M. I., Berndt, H., Chen, Y., Dixit, S., Jurva, R., Karppinen, P., Kekkonen, M., Kinnula, M., Kostakos, P., Lindberg, J., Mutafungwa, E., Ojutkangas, K., Rossi, E., Yrjölä, S., & Öörni, A. (Eds.). (2020). *White paper on 6G drivers and the UN SDGs. 6G Research Visions 2.* University of Oulu. http://urn.fi/urn:isbn:9789526226699..

Mejlgaard, N., Bouter, L M., Gaskell, G., Kavouras, P., Allum, N., Bendtsen, A.-K., Charitidis, C. A., Claesen, N., Dierickx, K., Domaradzka, A., Elizondo, A. R., Foeger, N., Hiney, M., Kaltenbrunner, W., Labib, K., Marušić, A., Sørensen, M. P., Ravn, T., Ščepanović, R., Tijdink, J. K., Veltri, G. A. (2020). Research integrity: Nine ways to move from talk to walk. *Nature, 586.* https://www.nature.com/articles/d41586-020-02847-8.

Microsoft. (2017a). Study: Why don't European girls like science or technology? https://news.microsoft.com/europe/features/dont-european-girls-like-science-technology/.

Microsoft. (2017b). Tutkimus: Miksi suomalaistytöt eivät kiinnostu luonnontieteistä? [Research: Why Finnish girls do not get interested on natural sciences?] https://news.microsoft.com/fi-fi/2017b/03/02/tutkimus-miksi-suomalaistytot-eivat-kiinnostu-luonnontieteista/.

Nilsen, M. W., Bloch, C. W., & Schiebinger, L. (2018). Making gender diversity work for scientific discovery and innovation. *Nature Human Behaviour, 2,* 10. https://doi.org/10.1038/s41562-018-0433-1.

Pursiainen, J., Muukkonen, H., Rusanen, J., & Sari, H. (2018). Lukion ainevalinnat ja tasa-arvo. [High School subject choices and gender equality.] No series. Oulu. http://jultika.oulu.fi/files/nbnfi-fe201803135965.pdf.

Saavalanien, P. (2020). Sustainability assessment tool for the design of new chemical processes. University of Oulu, Faculty of Technology. Ph.D. thesis. http://urn.fi/urn:isbn:9789526227719.

Sandberg, D. (2019). When women lead, firms win. S & P Global. https://www.spglobal.com/_division_assets/images/special-editorial/iif-2019/whenwomenlead_.pdf.

Schiebinger, L., Klinge, I., Sánchez de Madariaga, I., Paik, H. Y., Schraudner, M., & Stefanick, M. (2020). Gendered innovations in science, health and medicine, engineering, and environment. http://genderedinnovations.stanford.edu/.

The United Nations Educational, Scientific and Cultural Organization (UNESCO). (2017a). Cracking the code: Girls' and women's education in science, technology, engineering and mathematics (STEM). UNESCO. https://unesdoc.unesco.org/ark:/48223/pf0000253479.

The United Nations Educational, Scientific and Cultural Organization (UNESCO). (2017b). Measuring gender equality in science and engineering: The SAGA toolkit. SAGA working paper 2, UNESCO. https://unesdoc.unesco.org/ark:/48223/pf0000259766.

Smieszek, M. G., Prior, T., & Matthews, O. (2018). Women of the Arctic: Bridging policy, research and lived experience. Arctic Yearbook. https://arcticyearbook.com/arcticyearbook/2018/2018-briefing-notes/297-women-of-the-arctic-bridging-policy-researchlived-experience

Thomson, A., Palmén, R., Reidl, S., Barnard, S., Beranek, S., Dainty, A. R. J., & Hassan, T. M. (2021). Fostering collaborative approaches to gender equality interventions in higher education and research: The case of transnational and multi-institutional communities of practice. *Journal of Gender Studies.* https://doi.org/10.1080/09589236.2021.1935804.

Ulvinen, V. M., Vaara, H., & Kaleva, S. (2021). *Report on the best STE(A)M practices in Finland and in Oulu region.* University of Oulu, Faculty of Education. University of Oulu Study Materials Series, E18. http://jultika.oulu.fi/files/isbn9789526229928.pdf.

Chapter 4
Reflections on Selected Gender Equality in STEM Initiatives in an Irish University

Michael Devereux, Elizabeth Heffernan, Susan McKeever, Julie Dunne, Leslie Shoemaker, and Ciarán O'Leary

Abstract This chapter provides an account of the journey taken by the College of Sciences and Health in Technological University Dublin over an eleven-year period, from 2010 to 2021, as it sought to address the challenges of gender inequality in STEM (Science, Technology, Engineering, Mathematics) disciplines. The start and end points for the journey are marked by the formation of the College and its eventual replacement following a reorganisation of the Higher Education landscape in Ireland and a structural reorganisation at University level. This chapter draws upon the authors' collective experience of leadership within the managerial structures of the College and the University, and leadership on specific initiatives, to illustrate how consideration of gender equality and inequality featured within the journey taken by the College over this period of time. The chapter applies a broad lens to analyse the gender profile of the College's people and practices, including its planning, its operational management, its leadership, its staff profile and its student cohorts over the eleven-year period. The chapter also focusses more narrowly on a set of specific initiatives undertaken within the College or across the University which impacted upon the College. Collectively, the two perspectives demonstrate how a STEM College evolved over an extended period of time, shaped by cultural changes and challenges and bolstered by initiatives that targeted the areas of greatest challenge. This eleven-year snapshot provides insight into a journey that has built

M. Devereux · E. Heffernan · S. McKeever · J. Dunne · L. Shoemaker · C. O'Leary (✉)
Technological University Dublin, Dublin, Ireland
e-mail: ciaran.oleary@tudublin.ie

M. Devereux
e-mail: michael.devereux@tudublin.ie

E. Heffernan
e-mail: elizabeth.heffernan@tudublin.ie

S. McKeever
e-mail: susan.mckeever@tudublin.ie

J. Dunne
e-mail: julie.dunne@tudublin.ie

L. Shoemaker
e-mail: leslie.shoemaker@tudublin.ie

© The Author(s) 2022
F. J. García-Peñalvo et al. (eds.), *Women in STEM in Higher Education*, Lecture Notes in Educational Technology, https://doi.org/10.1007/978-981-19-1552-9_4

momentum and has potential to continue into the future. The story communicated in this chapter will be of value to Higher Education leaders and practitioners that wish to learn from this experience and interpret the approach set out in this chapter for their own local context.

Keywords Gender equality in STEM · Academic leadership and gender equality

4.1 Introduction

Ireland has evolved to become one of the world's leading hubs for life science and technology-based industries, leading to a growing demand for graduates with expertise and innovation capabilities in the STEM (Science, Technology, Engineering, Mathematics) fields. Skills shortages have been identified as the main inhibitor to growth and development in these sectors. For example, Engineers Ireland has recently reported that ninety-four per cent of relevant employers have cited skills shortages as the main barrier for growth (Engineers Ireland, 2021). This skills shortage is significantly influenced by a concurrent gender gap in STEM in terms of both enrolment on relevant University programmes and workforce profiles (Women in STEM Ireland, 2021). Furthermore, the pipeline of school leavers entering STEM programmes in higher education has been affected negatively by a significant decline in interest in STEM subjects for Irish secondary school terminal examinations (the Leaving Certificate), a situation that is affecting young women disproportionately (Department of Education, 2020). According to the Irish Central Statistics Office, women account for less than one-third of those employed in the Information and Communications Technology sector in Ireland and across the European Union (Central Statistics Office, 2019). In addition to a recognised gender pay-gap in STEM employment, gender imbalance is particularly pronounced in senior roles in industry where just seventeen per cent of Chief Executive Officer (CEO) roles are held by women (Women in STEM Ireland, 2021).

Higher Education has the potential to play a critical role in addressing gender inequality in STEM. However, Higher Education must first actively reflect upon how gender inequality is addressed in its own sector. In Irish Higher Education, for example, only twenty-three percent of professor roles are held by women (Higher Education Authority, 2018). Positive change in Higher Education requires Institutions to take action related to the gender profiles of their leaders, staff and students. Institutions need to critically reflect upon their organisational culture and nurture a culture that places equality of opportunity and experience for both staff and students at its core. This chapter provides a case study that describes the journey that the College of Sciences and Health in Technological University Dublin (TU Dublin) undertook over a period from 2010 to 2021. During this eleven-year period, the College has evolved towards a more inclusive and diverse working and learning environment that is internally and externally engaged in promoting women in STEM. While the College has experienced much success in this period, the journey is ongoing and the challenges of gender equality, and equality more generally, remain key concerns for STEM in TU Dublin. This is also the case for Irish Higher Education more broadly.

The chapter first provides a description of the College of Sciences and Health, TU Dublin, in Scct. 4.2. Sections 4.3, 4.4 and 4.5 deal specifically with staffing of the College and a critical evaluation of the changes that took place, or did not take place, during the period under consideration. This includes two sections titled 'Intervention in Focus', dealing specifically with the formation of a Women Leadership in Higher Education group in TU Dublin; and the provision by the University of gender training for researchers. Sections 4.6, 4.7, 4.8 and 4.9 address the student profile and student recruitment in the College. Three further 'Intervention in Focus' sections included in this part of the chapter address the formation by staff in the College of an Irish Network on Gender Equality in Computing Education; the development of an innovative mentoring programme in partnership with industry for women on STEM programmes; and the development—as part of the W-STEM project—of a workshop for second-level schools that challenges gender stereotyping of STEM roles.

4.2   The College of Sciences and Health, TU Dublin

As Ireland's capital city and largest city, Dublin has attracted substantial investment in recent decades from multinational companies, many of whom have established their European headquarters in Dublin. The city has become a European hub for the Information and Communications Technology (ICT) industry, and—along with other Irish cities—it has contributed to Ireland becoming an important European centre for other STEM industries such as Pharmaceuticals and Medical Devices (IDA Ireland, 2021). As such, there is a significant demand for talent in STEM fields in Dublin and throughout Ireland, a demand that has informed the shaping and strategy of Dublin's newest University, Technological University Dublin.

Technological University Dublin (TU Dublin) was established in January 2019 as the result of a formal merger between three Higher Education Institutions in the Dublin region: Dublin Institute of Technology (DIT), Institute of Technology Blanchardstown (ITB) and Institute of Technology Tallaght (ITT). The new University was the first Technological University to be established in Ireland following the publication by the Irish government of its National Strategy on Higher Education in 2011 (Higher Education Authority, 2011). Through this strategy, the government proposed the amalgamation of regional clusters of smaller institutions to form Technological Universities throughout Ireland, joining the community of similar Universities internationally that 'operate at the highest academic level in an environment that is specifically focused on technology and its application' (Higher Education Authority, 2011, p. 103). Through its preceding Institutions, TU Dublin has a rich history over one hundred and thirty years of providing technical and technological higher education in Dublin, with its graduates contributing nationally and internationally to social and economic development and culture and education.

Following its formation, the academic units in TU Dublin continued to operate in the preceding structure for a period lasting until August 2021, when the Colleges that had been established in Dublin Institute of Technology over a decade earlier were subsumed into new Faculties that spanned the full University. One of these Colleges, the College of Sciences and Health, had been established in March 2010 and was replaced in 2021 with the appointment of Deans for the University's new Faculties. This chapter uses the eleven-year period of existence of the College of Sciences and Health to bound a case study that explores how consideration of gender equality and inequality shaped its journey.

Throughout the period covered by this case study, the College of Sciences and Health was comprised of six schools in STEM disciplinary areas. The College had between 230 and 270 staff at any one time during this period, including academic staff, professional services staff, technical staff and management. The College was led by a Director and Dean (a single role) and managed through a College Leadership Team comprised of the Heads of School for the six schools, a College Manager, a Head of Learning Development and a Head of Research. Each school was managed by its respective Head of School and two or more Assistant Heads of School. Academic quality assurance for the College was managed through its College Board which in addition to including all management within the College also included staff and student representatives.

The College offered thirty programmes to school leavers in Ireland, attracting between eight hundred and nine hundred students onto these programmes annually. In addition, the College offered educational opportunities to mature learners, part-time learners and postgraduate students, and had a research student population of over two hundred students at any time. Overall, the College had a population of approximately four thousand students at any point in time. The College was also a primary contributor to research within both Dublin Institute of Technology and Technological University Dublin. From 2010 to 2021 the College was one of the largest providers of professional- and industry-oriented science graduates for the Agri-Food, Pharmaceuticals, ICT and Health sectors in Ireland.

4.3   Staffing, Recruitment and Promotion in the College

While the gender profile of academic staff in the College of Sciences and Health has remained consistent over the period being reported upon, with between forty-five and fifty-five per cent of academic staff from both genders at any point in time, the gender profile of senior positions in the College has seen a more substantial change in this period. As TU Dublin (and Dublin Institute of Technology prior to the formation of TU Dublin) is part of the Technological Higher Education Sector in Ireland, the academic staffing of the College is organised according to a linear progression route from Assistant Lecturer; to Lecturer; to Senior Lecturer I (a senior teaching position); to Senior Lecturer II (an academic management grade at which the roles of Assistant Head of School, Head of Learning Development and Head of

Research are positioned); and Senior Lecturer III (an academic management grade at which the role of Head of School is positioned). In 2010, 76% (n = 13) of the Senior Lecturer I positions in the College were held by male academic staff and 24% (n = 4) were held by female academic staff. By 2021, the gender profile at this grade had become 50% female (n = 8) and 50% male (n = 8). At management level in the College, the aggregated counts for staff at the Senior Lecturer II and Senior Lecturer III grades were 71% male and 29% female in 2010. By the end of the case study period in 2021, the gender profile at this grade had become 45% male (n = 8) and 55% female (n = 11). The College Leadership Team, described in the last section as having ten senior members from management grades in the College, has remained at 30% female (n = 3) and 70% male (n = 7) during this period of time.

The changes that have taken place in the College, or that have not yet taken place, are features of a journey that is ongoing. In 2010 the gender profile of the College was typical for an organisation that had concerning issues regarding gender equality and inequality. Although the overall gender profile across all staff was balanced (with greater than forty per cent of each gender represented) a significant imbalance was evident at senior academic and academic management roles. At this early stage it was recognised by the College leadership that there was frequently a dearth of applications for senior roles in general, with a particularly low level of interest being expressed by female staff in promotion to either Assistant Head of School (SLII) or Head of School (SLIII) roles. Through consultations with staff it became obvious that the culture of succession at that time was dominated by a perception that candidates were already identified for these positions prior to interview and in that context there was little value in competing for senior positions. In consulting with a range of female staff to try to identify roadblocks to applications, issues such as family commitments versus commitments of the roles, and other elements of the non-attractiveness of Head of School and Assistant Head of School roles also surfaced.

In the period from 2010 to 2021 the gender balance across management roles within the College has progressed in a positive direction, resulting in a viable pipeline of female applicants for senior roles within the new TU Dublin Faculties. The transition from the situation apparent in 2010 to that seen in 2021 was due to significant cultural and organisational changes. These included the cultivation of effective channels for communication for staff, students and other stakeholders and fostering an open and consultative culture in the College. In an open, trusting environment underpinned by inclusion and empowerment, the College sought to address the perceptions of staff regarding recruitment and promotion opportunities and encourage female staff to engage with the national- and University-wide supports that were being put in place (see Dunne et al., 2021 in this volume). In addition, transformative interventions and initiatives introduced within the College and TU Dublin more broadly have contributed to the progression of the College's approach to enhancing its gender equality profile during this period.

TU Dublin's human resources processes underwent significant change during this period, both for recruitment and performance management. Recruitment processes for all new appointments were transformed through the introduction of unconscious bias training for all managers who participate on recruitment selection panels; the introduction of mandatory gender balance on all interview panels; and the provision of effective feedback from interviews. The introduction of mandatory performance management and development for staff accompanied this change, leading to the development by staff of individual Personal Development Plans leading to Team Development Plans. Through this approach to planning, the College and the University were able to set and review targets that drew from individual experiences and goals. In 2018, Dublin Institute of Technology (of which the College was part at that time) was successful in its application for the Bronze Medal Award under the Athena SWAN programme (see Dunne et al., 2021 in this volume for further discussion regarding Athena SWAN). This award was made in recognition of commitment to advancing gender equality for women in STEM through organisational and cultural change.

Within the College, a new College committee structure was developed to engage and involve the wider College community in implementing the College strategy, and to encourage and facilitate inclusion, collaboration and sharing of resources. This committee structure acted as an effective channel of communication for staff, students and other stakeholders and fostered an open and consultative culture in the College. The College required that the membership of all College committees was gender-balanced, and set out this requirement in the terms of reference for the committees. The College introduced an expression of interest approach for membership of these Committees, replacing a nomination-based approach in which female staff were often under-represented.

Within the College, a number of influential female role models were recruited at senior levels. This included the appointment of one of the first female heads of a Computer Science school in Ireland, which ultimately transformed the culture and approach to equality, diversity and inclusion in that typically male-dominated discipline area. The School of Computer Science in TU Dublin has become a national and international award-winning leader in promoting women's participation in Computer Science. In 2020 the School of Computer Science was successful in attracting funding for one of twenty nationally dedicated senior academic Professors under the National Senior Academic Leadership initiative (SALi) (see Dunne et al., 2021 in this volume for further information regarding SALi). The funding facilitated the appointment of a permanent Professor of Inclusive Computer Science Education for TU Dublin.

The next two sections deal in detail with two interventions that had a significant impact on gender balance in the staffing and activities of the College of Sciences and Health in TU Dublin: the establishment of the Women Leaders in Higher Education group; and the roll-out across the University of gender-related training for researchers.

4.4   Intervention in Focus: Women Leadership in Higher Education

Since 2014, TU Dublin sponsored twelve women annually to participate in the AURORA Leadership Development Programme for Women in Higher Education (see Dunne et al., 2021 in this volume for further information regarding AURORA). A key message put forth in the AURORA programme relates to the benefit derived from being part of a strong network for women in higher education. This insight led to the formation of the Women Leadership in Higher Education (WLHE) network in TU Dublin.

A group of senior TU Dublin Managers from across the University, led by a senior academic manager in the College of Sciences and Health, established the Network in 2016 with the objective of encouraging and supporting women within TU Dublin to progress their careers and thus address imbalances in the gender profile at managerial level in TU Dublin and more broadly in Higher Education. The Network aims to be inclusive of women in academic, administration, research and technical roles at all levels and stages across TU Dublin. The Network hosts events, talks and workshops that are designed to encourage participants to take time for reflection on their personal and professional development and career advancement and enhancement.

To date the Network has held events in the thematic areas of:

- Professional enrichment, career advancement and promotion;
- Developing your professional supportive network;
- Being ready for opportunities;
- Leadership in higher education: where are the women?
- Transformation—the power of you;
- Conscious leadership—focus on the 'we' not the 'me'.

Keynote speakers have included female political leaders, educational leaders, television presenters and media experts, along with global business leaders. Speakers have shared their experience and personal journey to date including lessons learned and challenges along the way; their career launching platform (education/personal and professional supportive networks/mentors); how they made themselves ready to take opportunities and what influenced their choices; the values and beliefs they hold; and how they maintain a healthy work/life balance. More recently the Network launched a series of 'Come for Coffee' events, which aim to bring members of the Network across TU Dublin together to network with colleagues and to share experiences. These continued during the Covid-19 pandemic, as events moved to a virtual format. For its November 2020 event guest speakers explored experiences with Covid 19 and provided the audience with tools for supporting themselves, their families, colleagues and students. The success of the Women Leadership in Higher Education Network is evident in the high attendance to date, with each event attracting staff that recognise the need to focus on what they are doing to ensure that they achieve their personal goals.

4.5 Intervention in Focus: Researcher Gender Training

To ensure the gender element is considered within STEM research the main research funding organisations in Ireland, the Irish Research Council and Science Foundation Ireland, have included gender equality in their high-level strategic plans and consequent action plans. Each has introduced strategies to ensure that the research they fund meets the criteria that there will be gender equality in (1) research careers and teams and (2) integration of sex/gender analysis in research content (see (Dunne et al., 2021) in this volume for further information regarding these initiatives). To support its researchers in meeting these criteria the TU Dublin Research, Enterprise and Innovation Unit organised half-day researcher training workshops, one of which was hosted by the College of Sciences and Health in October 2019. The training was delivered by Yellow Window, a French-based global, multi-disciplinary product, service and public policy design agency employed by the EU to develop a Gender Research Toolkit and deliver training sessions all over Europe in order to give the research community practical tools to integrate gender aspects into research.

The objectives of the training workshop were:

- Strengthening participants' knowledge on gender issues;
- Raising participants' awareness on the importance of integrating gender in research;
- Informing participants on gender in Horizon 2020;
- Strengthening participants' capacities to integrate gender in research.

The resources were drawn from the EU Gender Toolkit, which comprises a series of documents available to download in English and Spanish (Yellow Window, 2012). The toolkit includes an introduction; a checklist for embedding gender in research; and a series of case studies and specific guidance and further reading for nine specific research fields: health; food, agriculture and biotechnology; nanosciences, materials and new production technologies; energy; environment; transport; socio-economic sciences and humanities; science in society and specific activities of international co-operation.

4.6 Student Recruitment and Support in the College

In the period under consideration in this case study, the College of Sciences and Health increased female student enrolments on full-time undergraduate programmes in the College by fifty-three per cent from 1,201 to 1,832, and increased the proportion of female students enrolled in those programmes from forty-five per cent to fifty-three per cent. Student recruitment in the College is undertaken in the constituent Schools (for example, see specific interventions designed by Schools in Sects. 4.7, 4.8 and 4.9) and on a shared cross-College platform. The College of Sciences and Health coordinated its cross-College outreach activities through its Public Engagement and

Recruitment Committee, a sub-committee of the College Leadership Team. It was a requirement that the committee, like all sub-committees of the College Leadership Team, consisted of at least forty per cent female and at least forty per cent male representation.

The outreach activities undertaken by the College included visits to secondary schools for career evenings, open days and class presentations. In addition, students from secondary schools visited TU Dublin's Science campuses on a number of occasions throughout the academic year, including a dedicated open day which took place in March or April of each year, a Science competition which invited secondary school students to submit a project, an open day for students from educationally disadvantaged backgrounds, an open day for students in Further Education Colleges, activities as part of National Science Week, Science camps and ad-hoc visits by schools. The College's Public Engagement and Recruitment Committee aimed to ensure gender balance in any situation where two or more staff were presenting to guests from secondary schools or while visiting secondary schools. The College implemented a Student Ambassador programme whereby students could apply to represent the College at public events. The College had built a requirement for representation of male and female students into the terms of reference for the Student Ambassador programme and included this as a statement on the application form. The College also ensured that all promotional literature presented an inclusive environment with diversity reflected among the staff and students identified by name or image.

The next three sections present specific interventions designed within Schools in the College of Sciences and Health.

4.7 Intervention in Focus: Gender Equality in Computing

It was one of my first labs—I stood at the door—there was already a big group of lads in there making a lot of noise—generally rowdy—and no other girls around. The lab assistant was sitting in the corner. I really wanted to turn around and run. I nearly did. But I went in and just sat quietly—too mortified to ask anyone how to get started.

—Female first-year Computer Science student, TU Dublin 2012

In 2012, the School of Computer Science in TU Dublin arranged a face-to-face meet-up of all female undergraduate students in order to allow this minority group to make connections and get to know other women in the school. Additional objectives were to gain insight into female students' motivations for selecting Computer Science as an area of study in Higher Education; exploring whether their experience was a positive one; and understanding how their experience could be improved. As a result of the findings from that meet-up, the School of Computer Science developed initiatives to increase enrolment of female students on its programmes. A key problem encountered was that while Computer Science departments in most Universities were putting real effort into gender balance initiatives in technology, each was working in isolation from others throughout the country, without sharing the best (and worst) practices that they were discovering. The School of Computer Science in TU Dublin (Dublin Institute Technology at that time) addressed this by setting up a collaborative network across Computer Science departments nationally. In 2017, they co-founded The Irish Network for Gender Equality in Computing (The INGENIC Network). This network includes a representative from each School of Computer Science (or Computing) across all Universities and Higher Education Institutions (HEI) in Ireland. The network is a simple but powerful way to enable HEIs to share information on their gender initiatives on the basic premise that working together is more powerful than working apart. Each representative has been able to highlight and share what they have learned on their gender balance journey, providing the implementation steps, costing, advice and benefits analysis. In addition to the benefit of shared learning, INGENIC has acted as a useful communication platform for data collection, with reach into every Computer Science department and Computing student in Ireland. INGENIC has also worked as a lobby group for change at government level, a tool for co-operation and information sharing, and a vehicle for attracting funding. In 2020, the Network attracted two funding grants to further develop the network as a mechanism for change in gender balance in Computing—and to develop a best practice toolkit for accelerating gender balance in HEIs. Representatives of the network meet three times per year, with hosting of the meet-up rotating around the members of the group. Communication is facilitated by an online repository for data and documents, and relevant email groups. Over time, peripheral benefits of the Network are emerging with wider research co-operation and strong working connections increasing across the individuals and Schools within the Network. The set-up cost of INGENIC is negligible, with each organisation committing to a representative being available for meet-ups and where possible, information gathering. The recent funding successes are now geared towards creating a more robust data sharing platform, and ongoing administration support for the network.

A second initiative established by the School of Computer Science is the provision of free coding (Computer Programming) workshops for female students from non-technology courses, including business, mathematics, biological sciences, law and social sciences. This initiative was run in co-operation with the CodeFirstGirls social enterprise organisation. In 2016, Dublin Institute of Technology (DIT) (now TU Dublin) became the first HEI in Ireland to partner with CodeFirstGirls to facilitate free coding courses. Between 2016 and 2019, more than 250 female students within DIT attended HTML and/or python courses, in most cases introducing them to coding for the first time.

The School of Computer Science implemented a seven-year programme, titled SUCCESS, to recruit and retain female academic staff (McKeever & Lillis, 2019). SUCCESS had a four-strand approach: Source, Career, Environment and Support. The Source strand explicitly encouraged women to apply for each recruitment drive; Career focused on female career and skills development initiatives; Environment created a female-friendly culture and reputation, and Support addressed practical supports for the specific difficulties experienced by female staff. In an academic team of approximately fifty-five full-time equivalents, thirty-six academic staff were female, and half of the senior academic leadership team were female, including a female Head of School.

4.8 Intervention in Focus: Educational Mentoring

The Equality in Science and Technology by Engaged Educational Mentoring (ESTeEM) initiative for female students was pioneered in September 2017. The development of the programme was shaped by research that demonstrated that when women are mentored by other women, in conjunction with other interventions such as gender-inclusive policies, there is a tangible and positive impact on their careers both for the short and long term (Lantz-Deaton et al., 2018). Mentoring is even more important where women are in work settings that are more male-dominated (Neal et al., 2013). Mentoring fosters support and co-operation rather than competition. Competition, which can be quite damaging, tends to be present in workplaces where the staff are significantly male (Chesler, 2009).

In 2015, the Organisation for Economic Co-operation and Development (OECD) published research about gender equality in education and STEM subjects which identified how adolescent girls lacked confidence in their abilities with STEM subjects (OECD, 2015). These findings were consistent with research Accenture conducted in Ireland and in the UK in 2015 (Accenture, 2015). The OECD report also proposed that some of these difficulties for young women in STEM subjects are due to a lack of female role models to emulate or look up to; and how a career in STEM does not fit into the stereotype of careers women 'should' be pursuing (OECD, 2015).

ESTeEM was designed to address these issues. Female students from Information and Communications Technology (ICT) programmes, Engineering programmes, and craft apprenticeships would be mentored by women from a related discipline in industry. As part of the mentoring, the student participants would develop a broader understanding about the range of career paths available as well as details about the skills required to be a successful STEM graduate. Since it is not uncommon for the numbers of female students in ICT, engineering and apprenticeship programmes to be low, which can also lead to a decreased retention of female students for a number of reasons, ESTeEM has additionally created an opportunity for the TU Dublin students to network with other students as well as professionals. Mentees are provided with a platform to meet other female students from other programmes which can lead to practical peer support while in third level. It is expected that these relationships will develop into professional connections during the students' future careers. Mentees are being afforded an opportunity to access a network of role models who are already established in their chosen careers; and learn from the insights and connections to prepare for the career opportunities and challenges that lie ahead.

ESTeEM targets two populations: the mentees and the mentors. For the mentors, the recruitment with the participating companies commences a month prior to the mentee's induction. The mentors are provided with a copy of the Mentor's Handbook and the Mentee's Handbook. These handbooks review the purpose of the programme and provide information about the range of programmes from which the mentees come. All mentors are invited to participate in a half-day of skills training prior to the first ESTeEM event, and all mentors are asked for details about themselves and their careers. This information is collated and provided to the participating students. This material enables the mentees to make an informed choice about their mentor while also increasing the student's 'buy-in' to the programme. Finally, the students who join ESTeEM are asked to provide information about themselves and this is provided to their mentor to assist with the first meeting and ice breaking.

The TU Dublin students are recruited for ESTeEM at the start of the academic year. The students who volunteer to join ESTeEM, including returning mentees, are invited to attend an induction where an overview of the programme is provided and the selection of mentors is initiated. At this meeting, they are provided with the Mentee Handbook and students are informed of the requirements, such as how they are expected to attend at least four of these events. Mentees who previously have been in ESTeEM are offered the option of choosing a new mentor. It has been found with this group of students, they value the opportunity to hear alternative perspectives and insights from a new mentor; and increase their professional network.

ESTeEM events are held as lunches during the academic year. Seated at each table are three mentors, their mentees (ideally each mentor has two students) and one TU Dublin staff member who is the moderator for the table. This set-up facilitates both networking and group discussions during the event, with eight to ten people at each table. The brief opening remarks start the event and then the main speaker, who is a person from one of the participating companies, discusses an agreed upon topic. The theme of the events, such as unconscious bias, networking or creating 'your brand', is kept broad and applicable to the student groups. At the end of each fifteen-minute talk, the moderator at each table will lead an activity with the mentors and mentees until the end of the event.

In 2019, when DIT received the Bronze Award for Gender Equity from Athena Swan, ESTeEM was selected as the example of best practice. Annually, the feedback via surveys to the mentors and mentees continues to be positive. In the words of a TU Dublin student:

> ESTeEM was one of my best experiences as a fresher [first year student]. Being shy didn't help me a lot but when I attend the first event it gave me the courage that I need. They helped me to come out of my shell, to experiment with new things and to create wonderful memories. My mentor wasn't just a person from a company, it was an entire table of students, lecturers and amazing people from industry.

—ESTeEM participant

4.9 Intervention in Focus: Challenging Gender Stereotypes

The College of Sciences and Health is a participant in the Erasmus + funded W-STEM project, through which it implemented a one-day workshop for second-level students to celebrate International Women's Day (IWD) 2020. The theme of IWD was #EachForEqual and the focus of the workshop called 'Let's talk about your dream job' was to raise awareness and challenge gender stereotyping and to consider how this can negatively affect the choices young people make about their future higher education and career planning.

The event, which was attended by seventy male and female students from three local secondary schools, commenced with activities to explore the students' attitudes to gender and stereotyping in various roles and careers. Many students continued to associate certain jobs with men or women. This was followed by a panel session, led by a TU Dublin Career Development Centre, with four professionals of gender non-typical to their field, including a female apprentice carpenter, a female apprentice plumber, a female computer scientist and a male primary school teacher. Speaking about their experiences, the panel demonstrated that gender is irrelevant in building skills in their chosen professions, and despite encountering challenges with bias, they spoke of the mentors that helped them along the way. Other activities included a talk about how lack of diversity in teams negatively affects innovation and design; Policies of inclusivity and the role of men in leading equality in Accenture; TU Dublin supports for diversity from the Students Union Vice-President for Student Welfare; and an advice session with TU Dublin Director of Equality, Diversity and Inclusion. The feedback received from attendees at the event was positive and highlighted the importance of students' biases and perceptions being challenged from a young age.

4.10 Discussion and Conclusions

During its eleven-year period of existence, the College of Sciences and Health in TU Dublin undertook a journey that helped it progress substantially along the path of equality, diversity and inclusion (EDI). Over these eleven years, the College addressed, to a substantial degree, the gender imbalance at senior levels of its academic staff and improved significantly the gender profile of its student cohort. The journey commenced by the College will continue into the future, as the EDI successes and ongoing challenges of the College may become a starting point for the new Faculties that are being developed in TU Dublin. This chapter has outlined several important aspects of the journey undertaken and highlighted a small number of influential initiatives whose design may serve as an indicator for Universities else-where of how interventions can successfully shape the gender profile of a University's staff and students.

A first, important lesson learned from this case study is the value of blending cultural and structural change. The everyday culture and lived experiences of staff and students, and the structures that form an organisation must both be captured by a lens that seeks to understand and address gender issues in an organisation. Changes to practice, or even changes to perception of practices, can have a significant impact and remove barriers to the satisfaction and progression of staff, and the options taken by students. Strong, focussed leadership from the management throughout the University is an important part of achieving cultural change.

A second lesson relates to the importance of building connections. The interventions detailed in this case study demonstrate the importance of connections being built through networks. Bringing people together to share experience and demonstrate support and leadership is an essential aspect of building a culture of belonging and inclusiveness. This applies in the case of networks being built internally within a University (such as the Women Leadership in Higher Education Network), across Universities (such as the Irish Network for Gender Equality in Computing) and between Universities and enterprise (such as the Equality in Science and Technology by Engaged Educational Mentoring programme).

The achievement of enhancement in equality, diversity and inclusion is a lengthy journey that needs to be nurtured and supported. The progression along this journey requires that academic leaders invest time in recognising the weaknesses within their organisation and address issues structurally and culturally. Elements of change require careful and thoughtful top-down leadership, but initiatives can also be created from staff who experience a supportive environment that is welcoming of change. Some of the most significant interventions in the College of Sciences and Health during the period under consideration arose from individual schools and individual staff taking responsibility for enacting change. The 'Interventions in Focus' sections in this chapter detail several occasions where this type of change was driven from individual experiences in the College of Sciences and Health. These successful initiatives can be replicated, built upon, learned from and enhanced in Universities elsewhere.

References

Accenture. (2015). *Continuing to power: Attracting more young women into Science and Technology 2.0*. Accenture.

Central Statistics Office. (2019). Women and Men in Ireland 2019. https://www.cso.ie/en/releasesa ndpublications/ep/p-wamii/womenandmeninireland2019/work/. Retrieved on 11 Oct, 2021.

Chesler, P. (2009). *Woman's inhumanity to woman*. Lawrence Hill Books.

Department of Education. (2020). Review of literature to identify a set of effective interventions for addressing gender balance in STEM in early years, primary and post-primary education settings. https://assets.gov.ie/96986/f05f7b2f-e175-442e-85e9-4a2264391843.pdf. Retrieved on 11 Oct, 2021.

Dunne, J, O'Reilly, A, O'Donoghue, A., & Kinahan, M. (2021). A review of Irish national strategy for gender equality in higher education 2010–2021. In this volume.

Engineers Ireland (2021) Engineering, 2021: A barometer of the profession in Ireland. https://www.engineersireland.ie/LinkClick.aspx?fileticket=wVvyHGDSRAQ%3d&resourceView=1. Retrieved on 11 Oct, 2021.

Higher Education Authority. (2011). National strategy for higher education to 2030. https://hea.ie/assets/uploads/2017/06/National-Strategy-for-Higher-Education-2030.pdf. Retrieved on 11 Oct, 2021.

Higher Education Authority. (2018). Higher education institutional staff profiles by gender. https://hea.ie/assets/uploads/2018/01/Higher-Education-Institutional-Staff-Profiles-by-Gender-2018.pdf. Retrieved on 11 Oct, 2021.

IDA Ireland. (2021). Facts about Ireland 2021. https://www.idaireland.com/newsroom/publicati ons/ida_facts_about_ireland_2021b. Retrieved on 11 Oct, 2021.

Lantz-Deaton, C, Tabassum, N., & McIntosh, B. (2018). Through the glass ceiling: is mentoring the way. *Internationa Journal of Human Resources Development and Management, 18*(3/4), 167–197.

McKeever, S., & Lillis, D. (2019). Gender equality initiatives and policies to develop the careers of female faculty; SUCCESS @ TU Dublin Computer Science. Informatic Europe. https://www. informatics-europe.org/component/phocadownload/category/16-develop-female-faculty.html

Neal, S., Boatman, J., & Miller, L. (2013). Women as mentors: Does she or doesn't she? Development Dimensions International, Inc, Pittsburgh. https://www.ddiworld.com/research/mentoring-women-in-the-workplace. Retrieved on 11 Oct, 2021.

OECD. (2015). *The ABC of gender equality: Aptitude, behaviour*. OECD Publishing.

Women in STEM Ireland. (2021). Women in STEM Ireland: Statistics and key findings. https://www. stemwomen.co.uk/blog/2021/02/women-in-stem-ireland-statistics-and-key-findings. Retrieved on 11 Oct, 2021.

Yellow Window. (2012). Gender in EU Funded research. Toolkit and training. https://www.yellow window.com/genderinresearch. Retrieved on 11 Oct, 2021.

Chapter 5
Balance4Better: "We Are HERe" More Than a Gender Campaign

Maria Giulia Ballatore, Claudia De Giorgi, Arianna Montorsi, and Anita Tabacco

Abstract Achieving gender equality and empowering all women and girls are part of the ultimate global challenge (Goal 5 of SDGs). The gender balance in STEM education is a challenge that has both horizontal and vertical dimensions. This paper focuses only on the first one. Politecnico di Torino, an Italian technical university with Engineering and Architectural courses, has a long history of attraction campaigns aiming to reduce the gender gap in its engineering enrolment. Despite these efforts, more remains to be done for the student population and high-level academic positions. During the academic year 2018/19, considering the engineering first-year enrolled students (around 4500), 25% were female, and a new innovative project was set, "WeAreHERe". This campaign aims to introduce a new vision to overcome both recruitment and retention: the female students become the main actors of the project by a guided training that let establish them as fresh role models. The use of social media and new technology support this storytelling and reach a variety of Italian girls. In this paper, the structure of "WeAreHERe" is described with some data analysis of its impact.

Keywords Attraction campaign · Gender gap · STEM education · Social communication

M. G. Ballatore (✉) · A. Tabacco
Department of Mathematical Sciences 'G.L. Lagrange', Politecnico Di Torino, Turin, Italy
e-mail: maria.ballatore@polito.it

A. Tabacco
e-mail: anita.tabacco@polito.it

C. De Giorgi
Department of Architecture and Design, Politecnico Di Torino, Turin, Italy
e-mail: claudia.degiorgi@polito.it

A. Montorsi
Department of Applied Science and Technology, Politecnico Di Torino, Turin, Italy
e-mail: arianna.montorsi@polito.it

© The Author(s) 2022
F. J. García-Peñalvo et al. (eds.), *Women in STEM in Higher Education*, Lecture Notes in Educational Technology, https://doi.org/10.1007/978-981-19-1552-9_5

5.1 Introduction

Higher education' attraction campaigns are a complex and structured mixture of activities and services devoted to recruitment students and serving the future job market. This play becomes even more crucial once it addresses STEM education. Here particular attention must be driven to equality, diversity, and inclusion. An inclusive attraction to STEM fields gives the basis to make the dream of a balanced future with equal opportunities come true. In this paper, we describe the "We are HERe" strategy that is in place at Politecnico di Torino, Italy starting from Spring 2019. This slogan is more than just an attraction campaign. It involves students and alumni directly and becomes the training set of young professionals to work together and experience the benefit of a balanced work environment. This paper aims to highlight the key feature of a successful experience reinforced by qualitative analysis. More precisely, we identify as crucial elements to evaluate the impact of the campaign the number of enrolled girls, the dropout rate, and the surveys' results.

Next paragraph pictures the political and socio-economical context we are in. We then describe the "We are HERe" project composition. The results and discussion support the theoretical approach used with observations and surveys results. In the end, we conclude by describing the future plans.

5.2 Context

State of the Art

One of the goals established to foster a more sustainable world is gender equality, as stated in goal 5 of SDGs: "Gender equality is not only a fundamental human right but a necessary foundation for a peaceful, prosperous and sustainable world" (United Nations, 2019). The gender balance in STEM is part of this challenge, and it has a double dimension: horizontal and vertical (Fulcher & Coyle, 2011). Scientists and politicians developed many different frameworks regarding the vertical balance to study what influences the career in STEM and how to support a balanced environment in academia (Bührer et al., 2019; Wolffram et al., 2017) and in the industry (Beede et al., 2011; González-González et al., 2018; Lambrecht & Tucker, 2019; Sassler et al., 2017). A large-scale longitudinal study in the USA found out that after 12 years from graduation, around 50% of women had left their job in the STEM field. Comparing these figures with the general one, the work shift cannot be linked to family factors.

In contrast, the work environment and the related job characteristics emerged as the key features (Glass et al., 2013). However, family factors play a role in the recruitment stage because STEM careers are perceived to fight with one's family goals (Weisgram & Diekman, 2015). This fact explains why having a woman family member in the STEM field favours girls' STEM interests (Cowgill et al., 2021).

Therefore, the horizontal gap represents the crucial "leaky pipeline" that needs to be overcome today to support future career changes.

Looking at the horizontal segregation, it is essential to distinguish recruitment from the retention campaign one. The first one refers to reinforcement on the attraction of young girls in the field, while the second aims to increase the retention of those already enrolled in a STEM and support their entrance into the labour market. Those are typically addressed with various strategies, and it is easy to mix the two campaigns. Steele has highlighted the importance of "rendering onto the right students the right intervention" (Steele & Aronson, 1995). For example, role models are valuable and effective in both moments. However, for the recruitment phase, it is important to have female role models (Makarova et al., 2019), while for retention, one can have both male and female role models to reach the desired goal (European Union, 2018). In recruitment, both the intrinsic and extrinsic motivations of the women concerning their academic decision have a dominant role. At the same time, in retention, the rationale mostly comes from personal stimulus and experiences. A review on the instruments used to study the gender gap in STEM stated that the main variables in these motivations are: influence established by this decision and the education path, the related achievements, the recommendations and work of parents, the stereotyped ideas they have towards this sector (Verdugo-Castro et al., 2019). Another factor that one must consider is the attitude towards STEM that can be measured by the spatial ability and, in particular, by the mental rotation factor, as shown in a longitudinal study of over 50 years (Wai et al., 2009).

How to translate all this input into a coherent and appropriately effective campaign? Many institutions have tried to answer this question through case studies (Garcia-Holgado et al., 2019; Politecnico di Torino, 2019), projects (Ballatore et al., 2020a; García-Holgado et al., 2020a), and events (Wyred, 2019). Currently, attraction campaigns use different media to foster a more balanced field, from the more traditionalist type (i.e., conferences, speech) to more interactive ones (i.e., summer schools, hackathons). Recently the technology has been used to spread the message among the youngest easily (García-Holgado et al., 2020b). Although the majority tend to forget the importance of increasing the retention of enrolled women, only a few experiences have been scientifically analysed (Gomez Soler et al., 2020). In general, the emphasis placed directly on the issue of the gender gap during the mediated strategy and, in particular, during events appears relevant. It is essential to make women feel welcome and not to emphasise the vertical dimension of the gender gap (ceiling effect, salary mismatch, and so on) in order not to have the opposite effect (Drury et al., 2011).

Local Context

Politecnico di Torino, an Italian technical university with Engineering and Architectural courses, has a long history of attraction campaigns aiming to reduce the gender gap in its engineering enrolment (Politecnico di Torino, 2002). The first woman who graduated in engineering in Italy was at Politecnico di Torino. Her name was Emma Strada, and it was in 1908. She was also the founder and the first president of AIDA, the Italian Association of Women Engineers and Architects. Despite these efforts, more remains to be done for the student population and high-level academic positions.

Considering the horizontal dimension, in its Strategic Plan "Polito4impact", presented in 2018, Politecnico has included a set of specific objectives to raise the average percentage of female students enrolled in the first year of engineering programs to over 35% and achieve full gender equality to some degree programs by 2024 (Politecnico di Torino, 2018). In that academic year (2018/19), considering the engineering first-year enrolled students (around 4600), about 25% were female, higher than the national one (23,8%). For further details, Figs. 5.3 and 5.4 report the historical series from 2015/16 to 2020/21.

Politecnico di Torino establishes a new vision to boost the girls' attraction. Instead of organising an unstructured series of events dedicated to young females, the idea is to have them as main actors. That is, regarding the retention, to create a female network of students in engineering that supports girls in career decisions through their testimony and the use of fresh content, proposed through media and new technologies. This shift will create "real" and "young" role models and ensure peer-to-peer interaction for the recruitment campaign.

5.3 "WeAreHERe" Campaign

History

The university established a completely new structure to reach the ambitious goals stated in Politecnico's strategic plan, the Equality@PoliTo. This new organisation fosters a new vision regarding horizontal segregation: the campaign should be more organic, address the target audience correctly and take advantage of social media. To do so, it seems essential to directly involve students in the creation of the new strategy. Therefore, on the occasion of the women's day (March 8th, 2019), a 24 h hackathon among our students was launched, SheHacks@Polito (Politecnico di Torino, 2019). This event was meant to convey the best suggestions for a campaign on these themes directly from the students' population. The winning project was "WeAreHERe", whose double meaning title perfectly reveals the purpose of the new desired vision. "We are her" and "we are here" at Politecnico, meaning that the best testimonials for enrolling new female students in Engineering are our female students themselves. This idea allows the recruitment actions to merge with the retention ones (see Table 5.1). By training our students, we reinforce our community allowing them to become the main actors as mentors of the recruitment stage and indirectly support them in a self-awareness path of the unique role they play in the field (retention phase).

Methodology

The attraction campaign focuses first on female students in the 14–18 age group. Several factors have suggested that such a group might be oriented towards STEM studies, but some fears hold back their choice. As an example, the analysis of the

Table 5.1 Overview of the campaign (HS: High-school students, BS: Bachelor students, MS: Master students)

		Audience	Main actor	Action
Contemporary role models				
One-to-one	Mentoring program	1-2y. BS	MS	Retention
	Study support	1y. BS	MS	Retention
	Calls	HS + 1y. BS	MS	Recruitment Retention
One-to-many	Daily storytelling on IG	Follower	3y. BS + MS	Recruitment Retention
	High-school Talk	HS	3y. BS + MS	Recruitment Retention
	Personality test	Follower	/	Recruitment Retention
Orientation support				
Incoming	Summer schools	HS	Experts	Recruitment
	"How to TIL"	HS	3y. BS + MS	Recruitment
	"Notte prima del TIL"	HS	3y. BS + MS	Recruitment
Outgoing	"AperiSTEM"	3y. BS + 2y. MS	Workers	Vertical gap
	Mentoring program	3y. BS + 2y. MS	Workers	Vertical gap
Experiences				
	Annual "e.vent"	HS + BS + MS	Guests	Recruitment Retention Vertical gap
	Annual "SheHackPoliTo"	BS + MS	BS + MS	Retention
	"We are HERe meets"	1y. BS	1y. BS + prof	Retention

results of the admission tests showed that even among those students who came within a step of enrolling, having passed the test, girls gave up completing the enrollment to a much greater extent than boys (30% more). The survey results we yearly propose to the young girls immediately after enrolling confirms the same suggestion. Looking at the data collected in fall 2020 (626 girls, 55% of the first-year engineering females, answered the survey), most respondents (48%) started to be interested in STEM at the age group 14–18. At the same time, 10% discovered this interest only recently, after 18 years old (Fig. 5.1). They identify themselves as the principal guide to their choice (64%). In particular, it is of interest that females recognise teachers as an even lesser guide (11%) than family (20%) (Fig. 5.2). The results confirmed the finding analysed in the literature review (Cowgill et al., 2021; Weisgram & Diekman, 2015) and suggested that much can be done to motivate high-school girls further to pursue engineering studies.

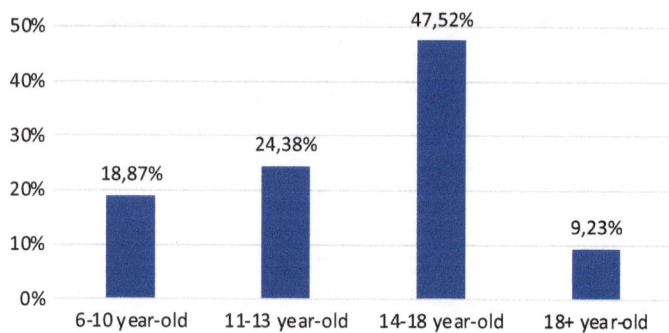

Fig. 5.1 Survey results related to the question "When did your interest toward STEM begin?"

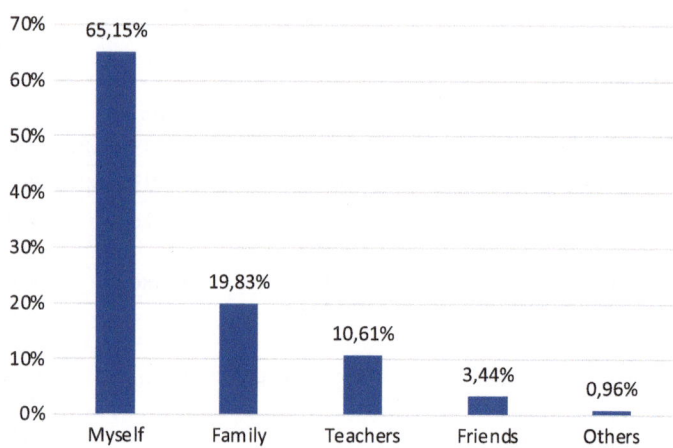

Fig. 5.2 Survey results related to the question "Who has most influenced your choice?"

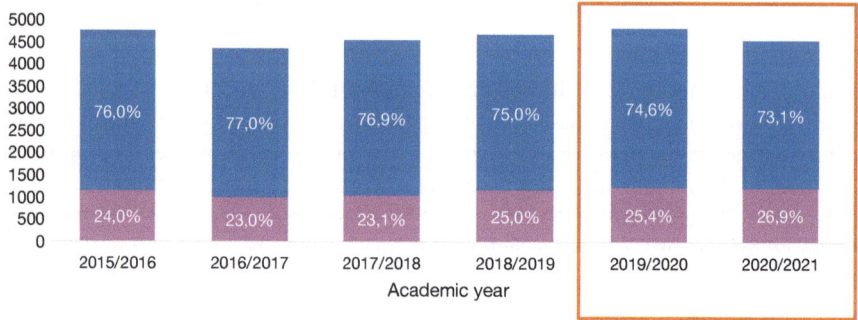

Fig. 5.3 Gender distribution from the academic year 2015/16 to 2020/21. Highlighted in red are the years in which WeAreHERe is in place

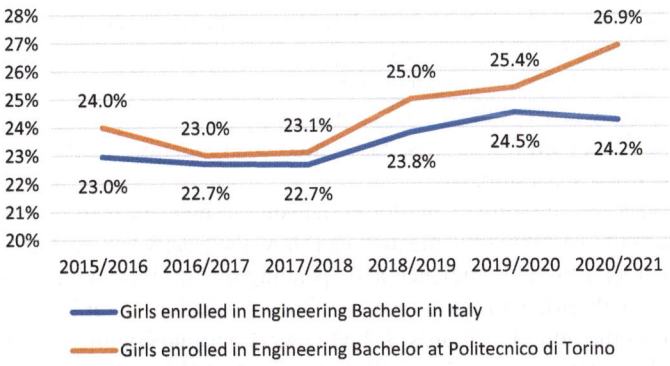

Fig. 5.4 Girls distribution from the academic year 2015/16 to 2020/21 at Politecnico di Torino and in Italy (*Data source* http://dati.ustat.miur.it/dataset/immatricolati)

The key element that characterises the attraction approach of WeAreHERe is to focus on experiences rather than formal events. By experiences, we mean events that require active participation with interaction and networking activities. We identified two main aspects to achieve this goal: the use of liquid communicative language and the transmission of content by other female students, perceived as peers. Through sharing their stories, via social media and meetings in high schools, as well as, in this period of COVID 19, through online events and one-to-one calls, the female students of the different engineering courses have become mentors for girls in high school. These young and familiar role models question with their testimony the stereotypes that represent one of the significant factors that prevent them from pursuing a STEM career. From the difficulty of the studies to the male environment, from the lack of free time to fearing a narrow work environment, one realises that what prevents girls from choosing this type of study often has nothing to do with their existing skills.

The difficulty in this approach is maintaining the freshness of peer-to-peer contact while providing informed content that avoids unconscious biases, like emphasising the gender gap and vertical segregation too much (Cowgill et al., 2021). University invites girls to become mentors through a general call open to all the Politecnico di Torino female students starting from the second year. Then they access special training by participating in a series of scheduled meetings. Initially, there are meetings with the Equality@Polito staff to raise awareness on gender equality, followed by a self-reflection on their experience to break the stereotypes. The second part of the training depends on their attitudes as the mentors are divided into two groups: one involved directly with high-school girls, while the other with first-year female students. In the latter case, the goal is to establish an empowerment community early in their engineering studies, which could help prevent the higher observed percentage of the first-year dropout among females (Politecnico di Torino, 2020).

Mentors meet with experienced mentors from the past edition and the external agency in charge of the overall campaign to be trained on typical questions and appropriate responses. To this end, an in-progress guide written along the first few years of the pilot is also available, collecting practical observations by students. Awareness-raising activities on stereotypes and unconscious biases are offered in the form of short inspirational videos, whose language aims to avoid pedantry. The campaign usually conveys the message through symbols rather than words: a female student on a skateboard whose wrist has a tattoo of an equation, a girl in her bedroom with a poster of a spaceship, etc.

WeAreHERe is a mixture of one-to-one and one-to-many peer-to-peer mentoring. Indeed, thanks to social media, mentors can show their daily life through Instagram stories reaching a broader audience and spending quality time with personal one-to-one chats with girls willing to receive more information. Satellite activities that work both on the attitude towards STEM careers and the cognitive side surround this storytelling: (i) summer schools (Ballatore et al., 2020b); (ii) interactive events; (iii) personality tests; (iv) study support; (v) career advisor. An annual event was established under the name "WeAreHERe e.vent" starting from 2020. It aims to create an appointment for all the people involved and interested in gender in STEM topics. For more information on how the different activities were structured, one can find references on both the website and Instagram channel of WeAreHERe (Politecnico di Torino, 2021).

Under a constructivist quarry, we evaluate this new methodology's impact on the target audience through a qualitative and quantitative approach (formal structured surveys and observations as well as an analysis of the students' interactions with AI tools, designed to detect their verbal impressions and elaborate them (Bethaz et al., 2020)).

5.4   Results

One can gain an insight into the results of the first two years of the project through both the enrollment data and the feedback collected in the annual surveys and the AI algorithms employed in the interactive events.

Since now, WeAreHERe has addressed Italian-speaking girls on social media and online activities, primarily local high schools for in-class talks. Therefore, we limit the population under consideration to this sample. In Fig. 5.3, we report the gender distribution relative to the first-year enrollment in engineering in the last six years. One can see that two years into the project WeAreHERe (from 2018/19 to 2020/21), there is an 8% relative increase in the percentage of engineering enrolment: from 25,0% to 26,9%. If the trend were to continue, by 2024, almost one-third of total engineering enrolment would be female.

Considering the national trend (Fig. 5.4), the results obtain by the campaign are even more evident. In the academic year 2020/21, the girls enrolled in an Italian Engineering course were 24,2%, −0,3% than the year before. This trend represents an opposite one compared to Politecnico di Torino, which recorded a + 1,5% increase. This data confirmed that the aim of WeareHERe is strongly linked to the Politecnico

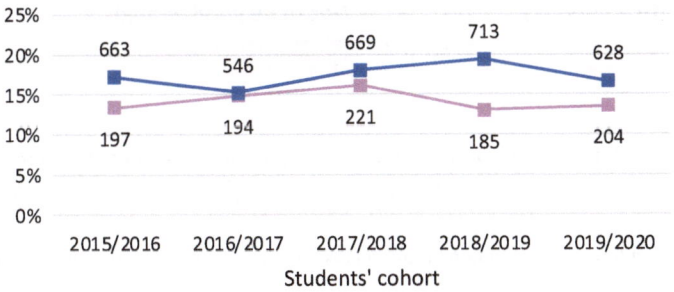

Fig. 5.5 End of first-year dropout rate for each cohort and gender (% on the total for gender)

di Torino enrollment as it fosters girls to STEM study and precisely to recruit them as future students.

Regarding retention, the dropout occurs mainly during the first year. Therefore, to have a clear picture, we consider the students that are still active at the beginning of their second year. Since the project started in March 2019, we can observe the first actions on the retention in the students' cohort 2018/2019. Before the project, the female dropout rates were about 15%, while now is about 13% (see Fig. 5.5).

Regarding the survey deserved to females right after enrolment, we compared the results of 2019/20 and 2020/21, finding the same trend. Therefore, in this discussion, we use the most recent data. The majority of the enrolled, 57%, have already heard, mainly via social media (about 40%) and institutional channels (about 40%). Of these, 20% considered the campaign a crucial factor in their choice. 78% of the enrolled female is interested in being supported by a graduate student (mentor) during the first year, mainly to receive support for the study, both in content and method, and socialising.

Social media, particularly Instagram, is a solid component of the project with about 3200 followers. Looking at the insights available on the app, the average stories' view is 1092, the average like at the posts is 149, and the accounts reached by at least one content in the last three months are 421.387.

The 2020 WeAreHERe e.vent was organised online and saw the participation of Sofia Viscardi, a writer, Youtuber, and influencer well known by the youngest (Politecnico di Torino and (YouTube channel), 2021). Around 5000 people attended the streaming, and until now, the video has achieved more than 30,000 views considering Youtube, WeAReHERe and Politecnico di Torino Instagram. A survey was sent to all the participants before the event, while we used an AI tool to analyse the trend topics. Figure 5.6 shows how young females perceive different conditions concerning their career decision. These results confirmed the literature and the survey analysis discussed above: personal motivation and fears are more critical than others' opinions. Figure 5.7 illustrates the words trends and the links between them. "To made" was the most used word with both meanings: creating something concrete and making new experiences. In (Bethaz et al., 2020), further details on the AI analysis of the event are given.

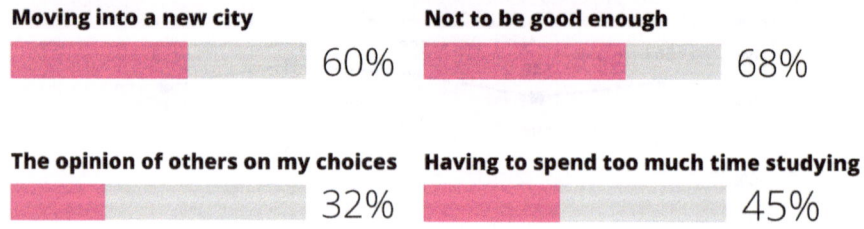

Moving into a new city 60%

Not to be good enough 68%

The opinion of others on my choices 32%

Having to spend too much time studying 45%

Fig. 5.6 Young females' perception of different conditions concerning their career decision

Fig. 5.7 Words trends and the links between them during the WeAreHERe e.vent 2020

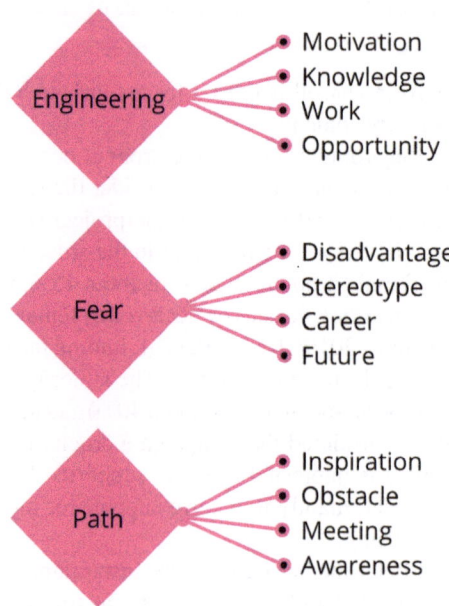

A further concrete byproduct is provided by the mentioned in-progress booklet prepared by mentors, which collects typical students' doubts and suggestions and is shared with the teaching office to improve the provided information to the whole students' community.

5.5 Conclusions and Future Plans

In the mid-term, the project's central goal is to train a multitude of skills of the female students who participate in the activities, such as creativity, scientific storytelling, curiosity, respect, and collaboration. They share their doubts and thoughts at large about Engineering studies with female students who overcame them. This personal growth translates into greater self-esteem. In the end, it envisages the project's self-sustainability in the formation of a generation of female engineering students aware of their choices, who will be able to motivate and be of reference for other female students in high school and beyond. In the long term, the project empowers the female component at Politecnico di Torino, increasing the gender equality and diversity of the whole academic community.

Covid-19 represented a challenge for the training and the in-person networking opportuning. However, thanks to the already in place multiple communication channels, we could maintain various activities both for the high-school population and university one and propose some new ones.

In conclusion, WeAreHERe positively impacted recruitment and helped reinforce a sense of community inside the female students. Nevertheless, we know that the 35% goal in the strategic plan is hard to reach by 2024. It is essential to highlight that this goal is very challenging. Any engineering university hardly reaches it in Europe, especially considering that Politecnico has an overall enrollment of 4600 students each year. A way to attain the goal is to extend the WeAreHERe attraction campaign to international students. This inclusion can be done by using both Italian and English for the social media content and having some dedicated mentors from countries different from Italy.

Regarding retention, to understand the added value of this campaign, we need to be able to analyse a more extended observation period. Right now, only the first-year dropout can be considered and not yet the entire three-year career. In general, knowing the crucial role of a gender-friendly environment, WeAreHERe will foster the whole students' population to believe in the importance of a more balanced world. We will achieve this future plan by promoting male role models that testify how together everything is more manageable and better: Balance4Better.

This positive experience has been shared with South American Universities within the "W-STEM: Engaging women into STEM, building the future of Latin America", a Capacity Building EU project. The use of fluid and fresh peer-to-peer communication to reach a wider audience, with the direct engagement of female students' as "role models" and mentors, is an innovative and creative way to attract and retain females in STEM careers that everyone can easily implement in other institutions worldwide.

References

Ballatore, M. G., De Borger, J., Misiewicz, J., & Tabacco, A. (2020a). ANNA tool: A way to connect future and past students in STEM. *IEEE Revista Iberoamericana De Tecnologias Del Aprendizaje, 15*(4), 344–351.

Ballatore, M. G., Duffy, G., Sorby, S., & Tabacco, A. (2020b). SAperI: Approaching gender gap using spatial ability training week in high-school context. In *Proceedings of the Eighth International Conference on Technological Ecosystems for Enhancing Multiculturality (TEEM'20)*. Association for Computing Machinery

Beede, D. N., Julian, T. A., Langdon, D., McKittrick, G., Khan, B., & Doms, M. E. (2011). Women in STEM: A gender gap to innovation. *Economics and Statistics Administration Issue Brief*, 04–11

Bethaz, P., Callà, R., Cerquitelli, T., Montorsi, A., & Giorgi, C. D. (2020). Proactive user engagement via friendly survey and data-driven methodologies. In *2020 IEEE 36th International Conference on Data Engineering Workshops (ICDEW)* (pp. 56–63).

Bührer, S., Reidl, S., Schmidt, E. K., Palmen, R., Striebing, C., & Groo, D. (2019). Evaluation framework for promoting gender equality in research and innovation: How does gender equality influence research and innovation outcomes and what implications can be derived for suitable evaluation approaches? *Journal for Research and Technology Policy Evaluation, 47*, 140–145.

Cowgill, C., Halper, L., Rios, K., & Crane, P. (2021). "Why so few?": Differential effects of framing the gender gap in STEM recruitment interventions. *Psychology of Women Quarterly, 45*(1), 61–78.

Drury, B. J., Siy, J. O., & Cheryan, S. (2011). When do female role models benefit women? *The Importance of Differentiating Recruitment from Retention in STEM, Psychological Inquiry, 22*(4), 265–269.

Fulcher, M., & Coyle, E. F. (2011). Breadwinner and caregiver: A cross-sectional analysis of children's and emerging adults' visions of their future family roles. *British Journal of Developmental Psychology, 29*, 330–346.

Garcia-Holgado, A., Vázquez-Ingelmo, A., Verdugo-Castro, S., González, C., Gómez, M. C. S., & Garcia-Peñalvo, F. J. (2019). Actions to promote diversity in engineering studies: A case study in a computer science degree. In *2019 IEEE Global Engineering Education Conference (EDUCON)* (pp. 793–800)

García-Holgado, A., Verdugo-Castro, S., González, C., Sánchez-Gómez, M. C., & García-Peñalvo, F. J. (2020a). European proposals to work in the gender gap in STEM: A systematic analysis. *IEEE Revista Iberoamericana de Tecnologias del Aprendizaje, 15*(3), 215–224.

García-Holgado, A., Verdugo-Castro, S., Sánchez-Gómez, M. C., & García-Peñalvo, F. J. (2020b). Facilitating access to the role models of women in STEM: W-STEM mobile app. In: Zaphiris, P., & Ioannou, A. (Eds.), *Learning and collaboration technologies. Designing, developing and deploying learning experiences. HCII 2020b.* Lecture notes in computer science, vol. 12205. Springer, Cham.

Glass, J. L., Sassler, S., Levitte, Y., & Michelmore, K. M. (2013). What's so special about STEM? A comparison of women's retention in STEM and professional occupations. *Social Forces, 92*(2), 723–756.

Gomez Soler, S. C., Abadía Alvarado, L. K., & Bernal Nisperuza, G. L. (2020). Women in STEM: Does college boost their performance? *Higher Education, 79*, 849–866.

González-González, C. S., García-Holgado, A., Martínez-Estévez, M. A., Gil, M., Martín-Fernandez, A., Marcos, A., Aranda C., & Gershon, T. S. (2018). Gender and engineering: Developing actions to encourage women in tech. In *IEEE Global Engineering Education Conference (EDUCON)* (pp. 2082–2087).

Lambrecht, A., & Tucker, C. (2019). Algorithmic bias? An empirical study of apparent gender-based discrimination in the display of STEM career ads. *Management Science, 65*(7), 2966–2981.

Liben, L. S., & Coyle, E. F. (2014). *Chapter three - developmental interventions to address the STEM gender gap: Exploring intended and unintended consequences.* Liben, L. S., & Bigler, R. S. (Eds.), Advances in child development and behavior, JAI (Vol. 47, pp. 77–115).

Makarova, E., Aeschlimann, B., & Herzog W. (2019). The gender gap in STEM fields: The impact of the gender stereotype of math and science on secondary students' career aspirations. *Frontiers in Education, 4*, 60.

Politecnico di Torino. (2002). Il progetto Donna: Professione Ingegnere https://didattica.polito.it/_progettodonna/progetto.html. Retrieved from 6 Oct, 2021.

Politecnico di Torino. (2018). *The Strategic Plan PoliTo4Impact.* Politecnico di Torino.

Politecnico di Torino. (2020). *Diversity is excellence: Gender equality report.* Politecnico di Torino.

Politecnico di Torino. (2019). She Hacks PoliTo. https://www.politocomunica.polito.it/press_room/comunicati/2019/she_hacks_polito. Retrieved from 6 Oct, 2021.

Politecnico di Torino. (2021). WeAreHERe. https://weareherepolito.it/. Retrieved from 6 Oct, 2021.

Politecnico di Torino (YouTube channel). (2021). WeAreHERe e.vent: un lavoro per Donne con Sofia Viscardi. https://www.youtube.com/watch?v=hpF-i-M0010. Retrieved from 6 Oct, 2021.

Sassler, S., Michelmore, K., & Smith, K. (2017). A tale of two majors: Explaining the gender gap in STEM employment among computer science and engineering degree holders. *Social Science, 6*(3), 69.

Steele, C. M., & Aronson, J. (1995). Stereotype threat and the intellectual test performance of African Americans. *Journal of Personality and Social Psychology, 69,* 797–811.

European Union. (2018). *Report on equality between women and men in the EU 2018.* Publications Office of the European Union.

United Nations. (2019). Goal 5: Achieve gender equality and empower all women and girls. https://unric.org/en/sdg-5/. Retrieved from 6 Oct, 2021.

Verdugo-Castro, S., García-Holgado, A., & Sánchez-Gómez, M. C. (2019). Analysis of instruments focused on gender gap in STEM education. In *Proceedings of the Seventh International Conference on Technological Ecosystems for Enhancing Multiculturality (TEEM'19)* (pp. 999–1006). Association for Computing Machinery.

Wai, J., Lubinski, D., & Benbow, C. P. (2009). Spatial ability for STEM domains: Aligning over 50 years of cumulative psychological knowledge solidifies its importance. *Journal of Educational Psychology, 101*(4), 817–835.

Weisgram, E., & Diekman, A. (2015). Family friendly STEM: Perspectives on recruiting and retaining women in STEM fields. *International Journal of Gender, Science and Technology, 8*(1), 38–45.

Wolffram, A., Aye, M., Apostolov, G., Andonova, S., O'Hagan, C., O'Connor, P., Chizzola, V., Çağlayan, H., Sağlamer, G., & Tan, M. G. (2017). *Perceptions of excellence in hiring processes: Results of mapping of the present situation in Bulgaria.* FESTA (Eds.), Sweden. https://www.festa-europa.eu/sites/festa-europa.eu/files/FESTA_5%201_final_report.pdf. Retrieved from 6 Oct, 2021.

Wyred. (2019). Online festival Wyred https://wyredproject.eu/online-festival/. Retrieved from 6 Oct, 2021.

Chapter 6
Promoting the Participation of Women in STEM: A Methodological View

Lucy García-Ramos, Rita Peñabaena-Niebles, Amparo Camacho, Maria Gabriela Calle, and Sofia García-Barreneche

Abstract The low participation of women in engineering fields is a common problem worldwide. As a result, universities are starting to create plans to attract more female students. However, there are no documented methodologies to guide institutions in this process. Hence, this chapter describes a method to attract more women to STEM programs, using one Latin American university as a case study. The procedure starts by establishing a baseline of the student population, using specific metrics to determine possible biases in admissions or graduations. The results show a small number of registered female students; thus, the method suggests different strategies to improve this situation. The next step is proposing activities to empower young girls to study engineering, describing indicators developed to evaluate the effectiveness of the activities. The case study includes participation from elementary, middle, and high school students. Results show that girls and boys participated in the proposed activities, and they maintained or improved their motivation to study a STEM program.

Keywords Gender · Equity · STEM · HIE · Methods · Institutional practices · Empowering

L. García-Ramos (✉) · R. Peñabaena-Niebles · A. Camacho · M. G. Calle · S. García-Barreneche
Universidad del Norte, Barranquilla, Colombia
e-mail: lucyr@uninorte.edu.co

R. Peñabaena-Niebles
e-mail: rpena@uninorte.edu.co

A. Camacho
e-mail: acamacho@uninorte.edu.co

M. G. Calle
e-mail: mcalle@uninorte.edu.co

S. García-Barreneche
e-mail: sbarreneche@uninorte.edu.co

© The Author(s) 2022
F. J. García-Peñalvo et al. (eds.), *Women in STEM in Higher Education*, Lecture Notes in Educational Technology, https://doi.org/10.1007/978-981-19-1552-9_6

6.1 Introduction

The low participation of women in STEM careers has been a problem for several decades. Reports illustrate this situation since the 1960s (Rossi, 1965; Goldman & Hewitt, 1976; Crowley 1977; Rossiter, 1993). Consequently, between 1966 and 1978, at least 300 projects were designed in the United States to increase the participation and status of women in science, engineering, and mathematics (Aldrich & Hall, 1980).

Furthermore, in recent years, the attention devoted to this situation has increased due to the relevance of science and technology in daily life and the search for solutions to contemporary problems. An example of this relevance is the increase in jobs in the STEM area compared to non-STEM areas, where growth is much higher in positions related to Computer Science (Zilberman & Ice, 2021). Additionally, STEM salaries are higher than other non-STEM careers (Funk & Parker, 2018).

STEM disciplines play a significant role in creating strategies, policies, and actions to achieve sustainable development. Thus, STEM education is vital at all levels, especially in primary and secondary education, to develop the capacities that future generations require to solve the complex problems they will face (Pahnke et al., 2019). However, despite this panorama, professionals in STEM careers are very scarce in some specialties such as Computer Science, Engineering, and Mathematics (NSF 2020).

Some studies try to determine the reasons for this phenomenon at secondary and tertiary education and entering the workforce (Cantillo & García, 2014). Other studies, rather than determining the causes of this situation, focus on developing strategies, projects, and actions to reverse it at the same levels (Charlesworth & Banaji, 2019; General Directorate of Communication Networks, Content & Technology of the European Commission, 2013).

Concerning the causes of this problem, (Charlesworth & Banaji, 2019) provides evidence on how gender differences in STEM careers have been explained based on individual characteristics, and specific differences in aptitude in mathematical competencies (Geary, 1996; Halpern, 1989). On the other hand, some authors argue that the historical gender divergences in average mathematical aptitude can be explained more by social barriers, especially considering the activation of gender stereotypes (Else-Quest et al., 2010; Gray et al., 2019; Hyde & Mertz, 2009; Penner & Paret, 2008). However, evidence from recent years indicates that the gender gap in performance in mathematics has been narrowing to reach tiny statistical differences (Feingold, 1988; Hyde, 2014, 2016; Lindberg et al., 2010; Zell et al., 2015). Thus, it is not a question of biological differences, but beliefs regarding mathematics, what influences performance in mathematical competencies (Hyde, 2016; Ceci et al., 2014). Women face barriers related to societal beliefs and expectations for women, especially in male-dominated STEM disciplines such as engineering, computer science, or the physical sciences (Swafford & Anderson, 2020).

On the contrary, there is increasing consensus that the factors related to the low participation of women in STEM careers are due to psychosocial and cultural factors such as gender roles, values, and lifestyle preferences (Ceci et al., 2014). These factors, in turn, are shaped by the sociocultural context in which people live and study. From early childhood, people receive social pressures about what roles are appropriate. For example, parents expect boys to be competitive, while girls prefer community and helping activities (Diekman et al., 2010; Eagly, 1987; Ferriman et al., 2009; Su et al., 2009; Weisgram et al., 2011). On the other hand, STEM careers seem like highly competitive and individualistic environments that value status and power. These particularities make women avoid these fields because they are contrary to female values (Diekman et al., 2011, 2015).

Sociocultural characteristics greatly influence all levels of society, specifically female high school students who want to start a STEM career. The underrepresentation of women in decision-making and STEM-related patents creates a bleak picture. It is common to see women in STEM fields in inferior positions, lower salaries, or earning the same wage as men but twice the responsibilities. These situations prevent improving the quality of employment for females in STEM areas (General Directorate of Communication Networks, Content & Technology of the European Commission, 2013; ECLAC, 2014). In other words, the imaginaries about women's performance in science, engineering, and mathematics make it difficult for women to start jobs in these fields (World Economic Forum, 2016).

Consequently, different studies proposed various models of inclusion and social representation of women in STEM fields to avoid prejudice or barriers (Aldrich & Hall, 1980; General Directorate of Communication Networks, Content & Technology of the European Commission, 2013). The initiatives start from preschool (Harris, 2020), primary (Milgram, 2011), secondary (Kang et al., 2019; Mosatche et al., 2013; Prieto-Rodriguez et al., 2020), and tertiary education at undergraduate (Cantillo & García, 2014; Smith et al., 2018) and graduate levels (Bekki et al., 2013). Other initiatives addressed the problem from the perspective of the organizational culture of higher education institutions (HEI) (Furst-Holloway & Miner, 2019) or as part of inclusion in the workforce (ASCE, 2021; Matthews nd). Other studies point to strategies, policies, or programs at the state level (Foundation, 2008) or directed to protecting the rights of underrepresented minorities in STEM careers (Syed & Chemers, 2011) such as the Afro-Latino population (1977; Morton & Parsons, 2018) indigenous (Stevens et al., 2016) or LGBTQI+ (Farrell et al., 2018; Yang et al., 2021), among others.

The recommendations for developing similar initiatives are diverse. For example, several experts emphasize the importance of collaboration between STEM programs at HEI and schools (González-Pérez et al., 2020; Milgram, 2011; Quigley et al., 2017). Additionally, some authors point to the importance of making role models visible (Herrmann et al., 2016; Math et al., 2012; Olsson & Martiny, 2018) to minimize the disagreement between the roles in STEM professions and sociocultural assumed appropriate roles for women.

Additionally, some ideas from the guidelines of General Directorate of Communication Networks, Content and Technology of the European Commission (2013) include using attractive STEM topics for young people and women in particular. These topics should be exciting, diverse, challenging, full of opportunities, and profitable. More ideas involve creating role models through the visibility of influential women in the sector, promoting a diversity approach and using inclusive terms, involving more men in the solution of this issue, and prioritizing STEM activities as not exclusive to one gender to avoid the bias.

The literature shows very different initiatives; however, there is no clarity in the methodological approach applied in these strategies. Thus, it is difficult to replicate them, creating limitations to solve the problem. Therefore, this chapter aims to describe the methodological approach developed during the execution of the project "Building the future of Latin America: engaging women into STEM (W-STEM), from here on called Project W-STEM. The project is funded by the European Union, under the ERASMUS + call: "Capacity Building in Higher Education Call for proposals EAC / A05 / 2017". The project aims to promote the participation of women in STEM careers. Section 6.2 shows a brief context of the W-STEM project. Then, the chapter illustrates the methodological approach in Sect. 6.3. Section 6.4 describes the execution of each stage in the Universidad del Norte. Section 6.5 shows the main lessons learned and the elements contributing to the project's sustainability. Finally, Sect. 6.6 concludes the chapter.

6.2 Brief Description of the W-STEM Project

The problem described in the previous section is a challenge to HEI in Latin America (LA). Thus, the situation leads to implementing a cooperation strategy among institutions in different regions. The goal of the ERASMUS+ call is to promote the development of capacities in HEI through transnational cooperation projects led by university networks. These projects strive to strengthen HEI's management, innovation, and internationalization capabilities. The primary strategy uses the funding priority "Equity, access and democratization of higher education" (García-Holgado et al., 2020).

The W-STEM project built a consortium of five European (Spain, Finland, Ireland, Italy, United Kingdom) and 10 LA universities (two for each country: Chile, Colombia, Costa Rica, Ecuador, Mexico). The University of Salamanca in Spain coordinates the project and interfaces between the EU Commission and the consortium. The project established the Universidad del Norte (Uninorte) in Colombia as support coordinator to facilitate management and communication between the LA partners. There is a steering committee with one representative for each partner, in which the project coordinator acts as a non-voting secretary.

The project includes different Work Packages (WP), and each one has a university leader and a co-leader. They assign tasks to the partners who participate in each WP and organize technical discussions, documents, and deliverables. The project also includes Columbus, an association for cooperation between universities in Europe and LA, as an external evaluator to guarantee the quality and success of the project.

The project has partners that receive no funding and play an essential role. One of these partners is UNESCO that supports national and international dissemination. Additionally, the participant LA universities have partners in high schools that support the campaigns for attracting and recruiting women to STEM careers.

The project established communication methods and instruments presented in periodic management meetings. The meetings allow for verifying the implementation of the action plans, preparing the respective minutes, and sending them to the entire consortium.

6.3 Methodological Approach

The methodological approach proposed in this section uses the conceptualization of the WP as a systematic process. Figure 6.1 shows that inputs and outputs aim to achieve the project's main objective. The figure illustrates four significant stages: definition of the baseline, information analysis, planning, and development. The results of each step constitute products that serve as input for the subsequent stage and allow for feedback and improvement of the process. The products related to the final stage define the project's results, including the analysis of indicators, lessons learned, and the sustainability of the actions. The following is a description of the methodological stages of the project.

6.3.1 Baseline

This stage determines the status of the member institutions regarding women's participation in STEM programs. The stage employs a quantitative self-assessment procedure and maps the critical processes of this study: attraction, access, advising, and retention. Furthermore, each university determines the starting point and defines strengths and weaknesses based on this diagnosis. Thus, universities define the opportunities for change that will lead to developing action plans to increase the participation of women in STEM careers.

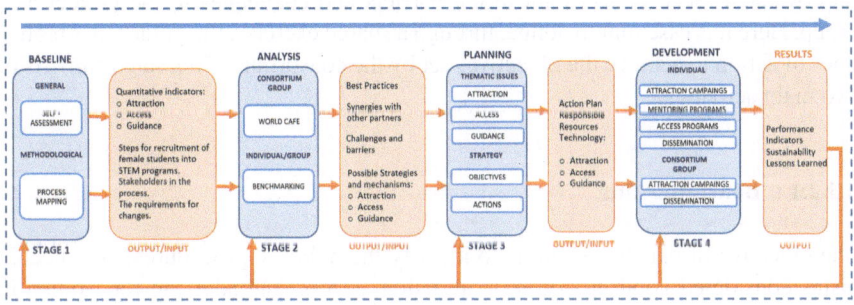

Fig. 6.1 Methodological approach

6.3.1.1   Self-Assessment

In this activity, each institution must collect all the quantitative information relevant to the processes under study to obtain data showing a gender gap. Therefore, the project designed an evaluation questionnaire (W-STEM Consortium, 2019a), based on a subset of the indicator matrix of the SAGA working document developed by Ernesto et al. (2018). The application of this questionnaire should result in the self-assessment matrix for each institution to help define the baseline indicators for each university and the consortium.

6.3.1.2   Mapping Process

This activity, which can develop in parallel with the self-assessment, aims at mapping the internal processes of attraction, access, advising, and retention of students in STEM programs. This mapping will help the institution determine all the steps involved in recruiting female students into STEM programs, the main stakeholders in the process, and the requirements for introducing changes.

For this purpose, the project designed a spreadsheet template (W-STEM Consortium, 2019b). Each institution must record the information associated with every process under study, defining the associated sub-processes, the activity description, and the responsible offices.

6.3.2   Analysis

After each institution determines the current state of women's participation in STEM programs, the institution analyzes the information collected to identify the gap, which actors are involved, and how to close it. This analysis considers two different scenarios. First, within each institution to identify strengths and opportunities for improvement. Second, to establish the barriers and challenges to the project as a group. Here it is essential to define, through a shared exercise, the guides to establish lines of action to achieve the objectives set by the consortium. This stage consists of two main activities.

6.3.2.1   Benchmarking

The objective of benchmarking is to identify the policies, procedures, and mechanisms considered good practices in attraction, access, and advising for women in undergraduate STEM programs so that other institutions can replicate them.

The project followed the methodology designed by Columbus for benchmarking clubs. Thus, the project created a questionnaire (W-STEM Consortium, 2019c) with four sections: description of the good practice, institutional strategies and policies, implementation and sustainability, and future improvements. After identifying the good practices, the institutions will socialize them in a group session.

The project has three Benchmarking sessions. The first one will be during the project's first year to socialize the good practices identified in the baseline stage. Additionally, this session will guide the elaboration of the action plan of the institutions. The other two benchmarking sessions will take place in the second and third years of the project.

6.3.2.2   World Café

This activity aims to identify policies, strategies, mechanisms, and actions to increase the participation of women in STEM programs. The World Café includes a collaborative dialogue and a group reflection with all consortium members. The institutions first share their experiences, then evaluate the main challenges faced in the processes of attraction, access, retention, and advising. Moreover, they define the ecosystem of strategies, mechanisms, and actions that aim to close the gender gap in STEM careers.

6.3.3   Planning

At this stage, each university defines the action plan to increase the participation of women in STEM programs. The institutions should explain actions to improve the processes of attraction, access, retention, and advising. Moreover, it is essential to ensure that these actions become policies to guarantee the sustainability of the objectives, even after the project ends. The action plans will adjust to the context of each participating institution, which will determine the mechanisms and how they will measure their achievements. For this purpose, the project designed a format to write the action plan for each consortium member.

6.3.4   Development

This stage executes all the mechanisms and actions defined in the planning stage. Baseline, Analysis and Planning occur in the project's first year. Thus, during the following two years, the institutions must develop the activities in the action plan, aiming to improve the processes of attraction, access, retention, and advising. Hence, specific actions as a consortium point to the design of attraction campaigns, mentoring programs, and other efforts to facilitate women's access to STEM careers. In addition, the project includes actions to disseminate the activities and the achievements obtained.

6.4   Engaging Women into STEM: The Case of Uninorte

The proposed methodology sets out a systematic process with guidelines, instruments, tools, and procedures whose application guarantees the achievement of objectives. Each institution seeks mechanisms and resources according to its context to implement the stages based on the proposed structure. This section presents a case study of how Uninorte approaches the execution and implementation of the W-STEM project from the previously described methodological approach.

6.4.1   Baseline

The definition of the current situation proposes implementing two activities that can execute simultaneously. The research group at Uninorte decided to perform them sequentially, starting with the self-assessment. After achieving significant progress, the group developed the process mapping. The main reason for this setup was that the first activity required more preparation and a more exhaustive review. The execution of these activities is explained below, showing the obtained results.

6.4.1.1   Self-Assessment

The first step in developing the self-assessment was defining the STEM programs participating in the study. In general terms, STEM programs stand between categories 5 and 7 of the "ISCED Classification" (Instituto de Estadística de la UNESCO, 2013). Therefore, Table 6.1 defines the undergraduate programs offered by the Uninorte with their respective ISCED code, name, and institution code.

The next step is to collect the information to fill out the self-assessment matrix proposed by the project consortium. The matrix has 26 sections to obtain quantitative indicators to measure the gender gap in the STEM programs for 2018 and 2019.

When analyzing the strategies for constructing the matrix, the first difficulty was that information was scattered in various university departments. Hence, the research group analyzed the different sections and questions of the matrix to identify 11 major categories of questions associated with almost 20 departments that could provide the required data. Then, the group requested the information from the graph of relationships related to the required information (see Table 6.2). At that time, another difficulty emerged because there were historical records of the information requested in most cases, but sometimes this information was not disaggregated for each program. Therefore, the group reprocessed the data, requiring more time and resources to complete the self-evaluation matrix. Some of the most relevant results of this activity are:

Table 6.1 Definition of STEM programs-Uninorte

Name ISCED	Code ISCED	Name Uninorte	Code Uninorte
Earth sciences	0532	Geology	GEO
Mathematics	0541	Mathematics	MAT
Electricity and Energy Engineering	0713	Electrical Engineering	IEL
Electronics and Automation Engineering	0714	Electronic engineering	IEN
Motor vehicles, ships, and aircraft Engineering	0716	Mechanical engineering	IME
Architecture and town planning	0731	Architecture and urban planning	ARQ
Building and civil engineering	0732	Civil and environmental engineering	ICI
–	–	Industrial Engineering	IIN
–	–	Systems Engineering	IST

Table 6.2 Relationships associated with the required information

Required information	Office	Required information	Office
Students and Alumni	Planning Office Alumni Office Academic programs	Mobility	OCI Academic programs Register Office
Attraction, access, and enrollment	Planning Office Admission Office Student Financing Office	Faculty	Faculty development Human resources Planning Office
Orientation, retention, and dropout	CREE Academic programs University wellness Register Office Planning Office	Student Financing	Student Financing Office Academic programs
Scholarships and financial aid	Student Financing Office Alliances and Corporate Relations	Awards and Recognition	Academic programs Register Office University wellness
Discrimination and sexual harassment	Human resources		

34.2% of the STEM population are women. IEL, IME, IST, and MAT programs have a women population smaller than 20%. The IEN program has female participation below 30%. STEM programs have a higher rate of male enrollees and admissions; thus, the situation is similar in the enrollment phase in seven of the nine STEM programs. Additionally, in six of the nine programs, the absorption rate (the ratio between enrolled and registered) is higher for men than for women. Nonetheless, the conditional probability of enrollment in a STEM program at Uninorte in 2018 does not depend upon gender. Therefore, the strategies to be implemented should be aimed mainly at the attraction mechanisms to raise the admission numbers to such programs.

Moreover, the process of access to an HEI in Colombia depends on external factors such as economic resources, obtaining scholarships, and admission to other Universities. Furthermore, the number of female professors teaching first-year STEM students is 26.7% of the teaching population at the university, which shows the lack of role models or references for girls in their first academic semester.

The achievements of this activity were first to establish the initial indicators that define the gender gap in STEM programs at Uninorte. Second, it created the need to conduct similar studies at least every two years to monitor the gap. Currently, the group is repeating this first step to obtaining the leading performance indicators for 2020 and 2021. Finally, although the university has a robust department that oversees the data and relevant information, the managers of these processes suggest handling centralized gender statistics allowing them to perform the data analysis more naturally and fluidly.

6.4.1.2   Mapping Process

The next activity at Uninorte was to map the processes associated with attraction, access, recruitment, retention, and advising developed within the institution. To this end, the research group designed a template to identify the sub-processes, activities, and responsible personnel. This process was simple because all the information resides in an easily accessible repository. Additionally, a flow chart documents the process and indicates every detail of the activities within each sub-process.

The process also identifies the stakeholders and their relationships. The responsibilities belong to a few departments such as Admissions, Marketing, The Office of the Registrar, Student Financing Office, academic programs, CREE (Spanish acronym for Center of Resources for Academic Success), and Student Affairs.

The research group identified robust processes for attraction, access and recruiting, and retention and advising. The institution has activities to strengthen the attraction processes, but there are no different activities for males and females.

The attraction processes encompass several mechanisms directed toward large high school populations, presenting all undergraduate programs differently. Nonetheless, the tools do not show elements to foster more extensive participation of women in STEM programs.

Even with all efforts in access and recruitment, the main obstacle to entering a STEM program is the cost of tuition. Thus, the group identified policies for obtaining scholarships and financial aid but did not find specific actions to increase women's participation.

The group also identified vital processes for retention and advising, but only related to obtaining good academic performance, regardless of gender. Additionally, the university provides workshops and other activities to foster professional skills in students. However, these activities do not consider the students' programs or their main gender problems.

The mapping process allowed the team to know the processes under study in detail. The section also defines the strengths and opportunities for improvement to create courses of action. Additionally, by reaching out to the people responsible for these processes, the team obtained support to develop activities related to the project. Thus, the team expects to achieve future sustainability to get longer impacts.

6.4.2   Analysis

This stage includes analyzing all information to define the actions for the following stages of the project. Therefore, each institution must analyze and socialize the results in the second consortium meeting. This meeting occurred before the end of the project's first year at the W-STEM International Leadership Summit. The stage was crucial to identifying the strengths and learning how to face the challenges. The following sub-sections describe the associated steps.

6.4.2.1   Benchmarking

Benchmarking includes identifying the good practices that can inspire other institutions. For this activity, the Uninorte team had previous experience as leader and moderator of U-Benchmarking Club: Improving the participation of women in STEM programs, organized by Columbus. The team already knew the methodology; thus, it could lead the first benchmarking round for the consortium.

Consequently, the team developed a form based upon the methodology created by Columbus. The document has four sections: description of the good practice, institutional strategies and policies, implementation and sustainability, and future improvements. The first section describes the type of practice and the expected goals. The second section shows if the practice results from an institutional policy or strategy; in this case, how the practice contributes to achieving the goals and what resources are available. The third section determines how to measure the success of the practice and the obtained results regarding learning and improvement. Finally, the last section describes how the institution visualizes the practice in the future and the actions planned to solve possible obstacles.

This document was socialized and sent to all the consortium universities; thus, they provided all information. The first round of the project benchmarking included presenting all this information at the W-STEM International Leadership Summit. A total of nine universities (seven LA and two European) presented their good practices during this session. Furthermore, another three (all European) presented them during the event aimed at university leaders. Finally, three universities (all in LA) expressed that they did not have good practices in these areas.

The good practice presented by the Uninorte team was the W-STEM Uninorte Student Group, a student community devised at the beginning of the project and founded in the summer of 2019 as a mechanism of attraction and retention of young women in STEM fields. The main objectives of this group are: creating a broadcast channel for the W-STEM project, attracting young women to STEM careers, providing pertinent information about STEM programs to young women in high schools, and raising awareness in the community about the importance of gender equity in STEM fields. Additionally, the student group develops the following activities: high-school attraction campaigns; robotics, programming, and data analysis hotbeds for college and high school students; conferences with local/national/international role models; tutoring programs for college and school students with former students and teachers; and activities where college students share their experiences as STEM students with high school girls.

As a result of this Benchmarking, the research group concluded that there is no statistical difference between LA and European universities regarding the participation of women in STEM programs. However, European Universities have a long way to go in studies and progress. Additionally, these countries have governmental and institutional policies that support these initiatives and actions. On the other hand, LA universities socialized robust initiatives that, in most cases, were isolated acts not supported by any institutional or governmental policy. Furthermore, this activity displayed good practices to replicate in different contexts.

The project conducted a second round of benchmarking in the second year. For this session, the Uninorte team presented the activities developed during the second year to promote the attraction of more women to STEM careers. The development stage will explain these activities in detail. Moreover, the consortium plans to do a final round to socialize good practices at the end of the project.

6.4.2.2   World Café

The second activity at the W-STEM International Leadership Summit was the "World Café", a group analysis that uses a brainstorming exercise to establish strategies and actions that the consortium institutions can develop to achieve the objectives from the Erasmus + W-STEM project. This activity follows a methodology based on sharing knowledge to establish possibilities for creative actions through a collaborative dialogue. Therefore, for guiding the conversation, the team prepared four tables with the following subjects:

1.   Public policies and institutional initiatives to promote the participation of women in STEM fields.
2.   Institutional policies and strategies to promote the participation of women in STEM fields.
3.   Strategies and mechanisms for ATTRACTION and ACCESS of young women to STEM careers.

4. Strategies and mechanisms for GUIDANCE, RETENTION, AND PROMO-TION of scientific careers for women.

The detailed description of the methodology used in the World Café and the main results are described in García Peñalvo et al. (2019). The Uninorte team reflected on and appropriated the results of this activity to plan the activities. Some of these are:

- The importance of

  - Developing long-term policies and plans requires creating mechanisms to include gender and STEM components in institutional policies.
  - Measuring the impact of the activities, analyzed, and executed with a gender perspective.
  - Raising awareness of the issue. Teach and prepare future leaders to be aware of the importance of having greater participation of women in STEM programs.
  - Reaching different stakeholders through social networks.

- Technological platforms can also help transform policies into actions, especially during the pandemic.
- Economic factors strongly impact women's access to STEM careers. Therefore, more targeted awards, scholarships, and long-term policies are needed.
- Teachers and families play a crucial role. They should be included in activities to promote women in STEM, to eliminate stereotypes and the perception that STEM fields are male-dominated.
- Creation of Groups/Communities and support groups for women in STEM, so members participate in seminars, talks, and gender awareness meetings.

6.4.3 Strategic Planning Process

For developing action plans aimed at the attraction, access, retention, and orientation processes, Uninorte included individual and collective information analysis. The analysis encompassed all WPs such as the Self-Assessment on Gender Equality in STEM, the Process Mapping of Attraction, Access and Guidance, the Application of Self-Assessment Tools, the Analysis report, and the Benchmarking rounds on strategies and mechanisms.

For planning, the team employed the methodology established by the project to formulate the plan, including creating improvement objectives and the actions for achieving them. The first step was to analyze the current state of the three processes in the institution and discuss macro strategies that served as inspiration for the formulation of the objectives. The second step was defining actions to achieve the goals, displaying them in a matrix that mapped the activities to the purposes. Finally, the team assigned the people responsible for each step.

The research group formulated the action plan to achieve the following goals focused on STEM programs.

Attraction: To increase the number of female students, the team established the following actions.

- Develop awareness days with directors of partner schools about the problem of participation of women in STEM careers to obtain a commitment to support the campaigns to be carried out within their institutions.
- Develop a training session with vocational counselors and science and technology teachers from partner schools to show STEM careers and their importance in the development of society.
- Design and implement campaigns aimed at senior students at the partner schools, particularly women, to recognize STEM programs as options for university study and professional life.
- Access: To create mechanisms for assuring recruitment, the actions were:
- Hold meetings with administrative offices (admissions, graduates, fundraising, among others) to discuss, reflect, and propose strategies to facilitate and support women's access.
- Analyze the structure required (financial and operational) to propose a scholarship in STEM undergraduate programs aimed at candidates who need funding.

Retention and Guiding: to establish mechanisms for monitoring and to advise students from the first year, including:

- Develop a focus group with students of the primary cycle to know the main academic barriers for their student success.
- Design and implement an activity to support students' success in their basic cycle.

6.4.4 Development

6.4.4.1 Attraction and Recruitment Campaigns

According to the methodological approach, the next stage consisted of developing the attraction campaigns following two principles according to the literature: Joint university-school work and visualization of female STEM role models.

The target population of the attraction campaigns was senior secondary education students in the region, so they know the women who work in STEM and their contributions to the industry, academia, and civil society (García-Ramos et al., 2021). Additionally, the campaigns provided information on the different possibilities for women when choosing a STEM career and the skills and abilities required. Although the attraction campaigns were designed for girls, due to the COVID pandemic, the participation was open to all students. The main campaign subjects were W-STEM Cinema, Computer Workshops, and Role Models in STEM.

W-STEM Cinema

The first activity consisted of a cinema forum; a space created to reflect on the role of women in STEM through the film industry. The sessions included:

Hidden Figures: This activity highlighted the qualities and attitudes of the lead characters, which allowed them to succeed despite the double discrimination in the 60 s.

Shakuntala Devi, "The Human–Computer": This work recounts the exciting life of the Indian mathematician and writer who, without any formal education, achieved multiple feats by performing mathematical calculations.

Let's talk about the girls of The Big Bang Theory: The stereotypes about men and women were analyzed, especially the evolution of the central female character who starts as the stereotype of a dumb blonde girl and later becomes an empowered woman and friend of female scientists.

Interstellar: the activity analyzed the protagonism of science and technology and the varied and positive female roles.

Programming Workshops

The research group organized several workshops to motivate students to develop and strengthen their computational skills because they are essential in any work, especially in STEM fields. The workshops were:

Code.org: the activity presented the Code programming learning platform that uses a didactic and playful format to fulfill the educational needs of primary and secondary students.

PSeInt: this workshop introduces this educational tool, very popular in universities in LA and Spain, to learn programming fundamentals in a pseudo interpreter.

Java: this seminar introduced the programming language and its importance in the software industry through didactic and straightforward examples.

Python + Colab: this workshop introduces the Python programming language on the Colaboratory platform.

Role models and empowerment

One good practice was to promote empowerment through role models in STEM to show students a greater diversity of a woman's roles. In addition, this campaign aims to empower STEM women in their different fields of action. The campaign includes the following presentations:

Investigating Biomass energy to mitigate climate change: The presenter was a female researcher with the Biomass Energy with Carbon Capture and Storage (BECCS) line of the Tyndall Centre for Climate Change Research (United Kingdom). She shared the results of her doctoral thesis.

Renewable Energies, management of a photovoltaic solar energy project: a young female engineer presented this webinar, showing her role as manager in constructing a self-generating solar farm.

From girl to engineer: This activity consisted of a series of talks to highlight the work of several award-winning STEM women in their areas. Each woman presented her life stories from childhood to her professional life, showing the skills that allowed her to achieve her goals.

Women in Engineering W-STEM + WIE: Four female engineers presented their perspectives and experience to high-school and university students.

Colombian Women in STEM: This activity invites Colombian women who stand out internationally to share their STEM experiences with the viewers.

STEM and its magical world: the group organized activities exclusively for high-school students as the schools gradually resumed their presence in the classroom. In this case, senior students from STEM programs talked to school students about STEM careers.

Monitoring of activities

To keep track of the attendance at each event, participants had to register before starting each activity, and they answered one question about their interest in studying a STEM program. Then, at the end of some events, the group asked attendees the same question to determine if the activities changed the students' perception. Table 6.3, Figs. 6.1 and 6.2 show the participation achieved in the campaigns and the students' perception before the event.

Table 6.3 Registered and participated students in the attraction campaigns

Campaign	Students register	Students participant	STEM careers interest before Total responses
Cinema W-STEM	85	53	63
Programming workshops	557	303	530
Role models and empowerment	179	116	133
Total	821	472	726

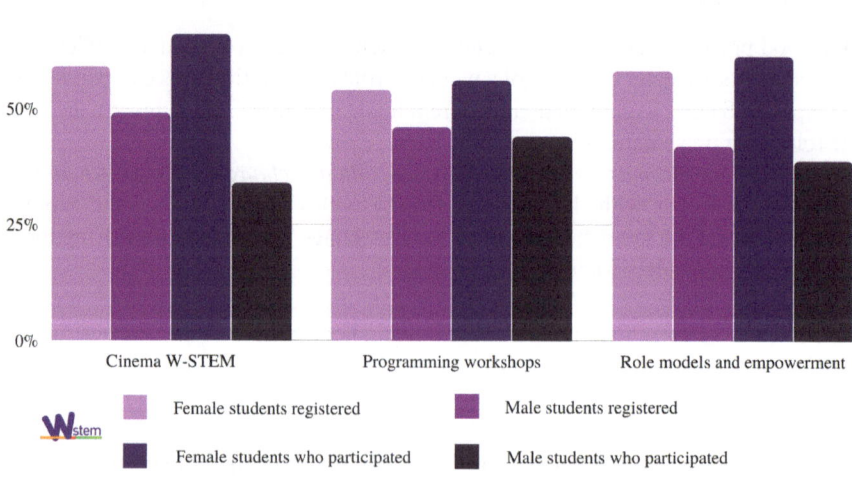

Fig. 6.2 Students registered and participated in the attraction campaigns

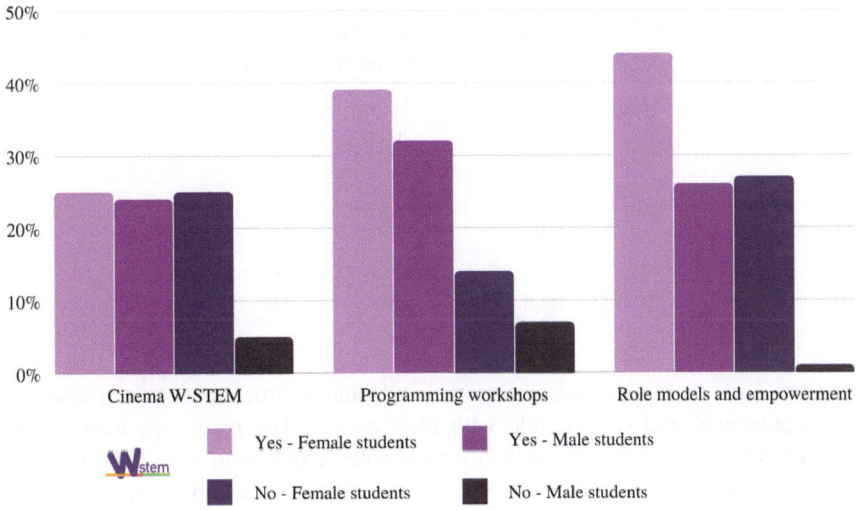

Fig. 6.3 STEM careers interest before the attraction campaigns

Table 6.4 Perception of the students post attraction campaigns

	STEM careers interest after						
	Total	Yes		Not sure		No	
Female	69	41	59,42%	16	23,19%	12	17,39%
Male	16	15	93,75%	1	6,25%	0	0%

Table 6.3 shows that 58% of the participants are women. Additionally, Figs. 6.2 and 6.3 show that 39% of women and 30% of men stated they had considered studying a STEM program. However, 5% of the male attendees said they were not interested in a STEM program, in contrast to women, whose responses reached 17%. This situation may reflect a perception influenced by gender stereotypes in STEM careers, where the possibility of choosing a profession of this type is perceived as more appropriate for men.

Finally, Table 6.4 presents the survey results at the end of some events. Unfortunately, due to logistical issues, the group could not apply it in all the events, nor did all the students respond. Nonetheless, Table 6.5 shows that the percentage of women who showed interest is 59%, while the ratio of men is 94%. The result is encouraging because it offers a surprisingly high rate for female students.

6.4.4.2   MentorADA

The Uninorte team created the MentorADA mentoring program to form a support network for guidance and advice for students of STEM careers. Additionally, the

Table 6.5 Comparison of program students and registered students for MentorADA

Program	Total First semester	Female First semester	MentorADA First-year
IEL	5	2	3
IEN	20	4	5
MAT and Data Science	13	5	3
IME	23	1	6
IST	43	10	5
Total	104	22	22

program aims to support students in their adaptation process to the university to feel represented and comfortable with their choice. The program's name maintains the principle of making visible female role models who have contributed to STEM, particularly as a tribute to the mathematician Ada Lovelace, a visionary of computer programming (Carlucci Aiello, 2016). The MentorADA program is directed to students from the STEM programs with lower participation of women and includes the following stages.

Team building

The first step was to define the target population of the MentorADA program, first-year female students from STEM programs with less female participation: IST, IME, IEL, and IEN. MentorADA also included a recently created STEM program: MAT and Data Science.

The second step was to identify outstanding students, professors, and graduates in the STEM disciplines and send invitations to participate in the mentoring group. Currently, the group has four mentors (professors and alumni) and 14 senior mentors (senior undergraduate students). The MentorADA team includes 21 people counting the W-STEM project staff.

Planification of MentorADA

The MentorADA team devised the structure of the network and the role that each member would play. For this purpose, one mentor per program should provide a global vision and create spaces for learning specific aspects of the program and the short- and medium-term skills required. The mentor should also identify possible misconceptions that may hinder the mentees' performance.

Additionally, the senior mentors will accompany the students to integrate into the university. These are senior students with good academic performance and working with various student groups. The senior mentors would also support the execution of specific activities of the mentors and provide personalized follow-up to the mentees assigned to them.

Fig. 6.4 First semester students from STEM programs

82 male students

22 female students

Training

After defining the team structure, functions, and program guidelines, the team set up training workshops for mentors and project staff. Professionals from Student Affairs at Uninorte conducted the workshops developing the most critical topics to allow mentors to perform their work in the best way. The training included four thematic axes: mentoring, emotions, empowerment, and leadership.

Preparation and execution of the action plan

Based on the established guidelines, the training, and their own experience, each mentor developed their work plan, reviewed and approved by the project staff. Next, the staff opened enrollment for first-year students of the identified programs. To disseminate the call for mentees, the team used mass mailings and posts on social media accounts. Furthermore, the team selected the first-semester course of each engineering program and visited the class to inform students about the MentorADA program and its benefits. Table 6.5 and Fig. 6.4 show the data of new students in addition to those who joined the MentorADA program. The table shows that the MentorADA program attracted most first-year female students.

6.4.4.3 Activities to Support Access to STEM Programs

The initiatives that best contribute to the access of young women to STEM careers at Uninorte are:

The creation of the Marvel Moreno scholarship to honor Barranquilla writer and outstanding woman in Colombian literature. The scholarship aims at high school women from the Colombian Caribbean region with exceptional academic performance during their secondary education. The applicants must be interested in studying MAT, ICI, IST, IME, IEL, and IEN, among other programs. The Marvel Moreno scholarship applies to young people of any socioeconomic level, preferably with economic difficulties to access university.

The structuring of a new scholarship, to be called W-STEM. The Uninorte team is currently studying this scholarship. The idea is to expand the opportunities for access exclusively to STEM careers for young people with economic difficulties but with outstanding academic performance in high school and motivation to study these programs.

6.4.4.4   Dissemination

The student group constantly carries out dissemination activities open to the university community and other interested parties. These activities deal with issues of gender equality to construct a more equitable society. Additionally, the activities create spaces for students to develop their soft skills and provide them with content informing them on gender issues.

This semester the group focused on using social networks (especially Instagram) to make more attraction and dissemination campaigns. The main reason was that the group could reach many more students in the region with graphic and audio-visual tools or content. Through the W-STEM Uninorte account, the group shared illustrative posts about pioneer women who have changed the world. The group also addressed related topics and highlighted the importance of diversity and female participation in STEM. This account also shares announcements of scholarships and national and international opportunities in STEM fields.

Additionally, the group recommends books and movies where the lead characters are women in STEM. Furthermore, the group provides tools that can be useful in their path as STEM students in college. Another type of dissemination activity is workshops, ranging from programming and software management to emotional intelligence and leadership.

The group has become a support network for students to share their experiences and learning. In case of needing professional help, the team makes the referral to the relevant entities of the university, such as the team of psychologists of Student Affairs and the academic program coordinator.

6.5   Discussion

6.5.1   Lessons Learned

The first big lesson is to value the importance of data and information as a measure of performance and analysis to detect the gap and identify the mechanisms to close it. This lesson helped establish appropriate indicators to measure the processes from a gender perspective. Thus, the research group took the first steps to start measuring and analyzing data to develop a gender equality policy, starting with STEM programs but reaching all institution levels.

The second lesson learned was that the research group is not alone. The best results are achieved by involving those responsible for the processes under study and showing the benefits of approaching their activities from a gender perspective. Thus, identifying internal strategic partners to support the project's development and to become future managers of the activities allows the sustainability of policies to close the gender gap in STEM programs over time.

Another lesson is the importance of networks or working communities to reflect and share experiences adapted to different contexts.

A further lesson learned is the importance of obtaining the support and commitment of institutional leaders to enable the implementation of more initiatives and obtain the necessary financial aid. One strategy to get this support is raising awareness in the institutional leadership about the problem. Additionally, the leadership must know about the benefits of increasing the population of women in STEM areas to the institution, the city, the region, and the country. Finally, it is also possible to seek strategic business allies for financial support.

A possible bias of the study is related to the profile of the participating students: did they attend the activities because they already had a prior interest? Did the project manager reach students who did not consider STEM careers among their options? Did it change their perception of these disciplines? All these questions should be answered with the evaluation instruments; however, the post-activity response rates were meager; therefore, no generalizations can be made about the results. To deal with this situation, strict planning of the activities should include a time margin to answer the evaluation instruments as part of the activity developed.

6.5.2 Sustainability

The project's sustainability relates to an institutional culture of support and encouragement for the access and permanence of young women in STEM programs. This culture employs policies, strategies, and procedures for attraction, recruitment, admission, advising, and permanence.

Uninorte has formulated strategies and developed initiatives and activities for these processes. The most important of them are:

The attraction, dissemination, and mentoring campaigns motivate young people from schools, Uninorte, and other universities in the region and the country. Every day more and more young people are involved in these campaigns.

Institutional scholarships, such as Marvel Moreno and the future W-STEM scholarship.

Creation and consolidation of the W-STEM student group, which serves as a model for the involvement and commitment of young women in these initiatives.

With similar initiatives and activities incorporated into the daily routine of academic life, the W-STEM project becomes an integral part of the institution's educational offerings, guaranteeing the project's sustainability.

6.5.3 Limitations

One of the main limitations in carrying out the activities in the schools was the reduced response of school principals to the calls. With the abrupt changes in teaching modalities during Covid confinement, some school principals claimed to be overwhelmed with the new demands and therefore declined to participate. Additionally, students also suffered consequences of these abrupt changes, aggravated by the limited connectivity, even within the urban area. This situation defined the activities' schedules; initially, they were planned in the mornings within the school schedule, but student participation was meager. The results could be very different if the project included activities with the teachers who directly contact the students. Undoubtedly, achieving teacher engagement will increase the impact of the project.

Due to the low response rate from school principals, the project staff used other communication channels such as contacts from the most used social networks such as Instagram and networks of friends through Whatsapp. Although these networks were very effective in some cases, there is a possible bias in the participants as they come from a direct or indirect relationship with the project staff. The challenge, in this case, is to make broader calls that include a variety of student profiles.

Another limitation in developing the attraction campaigns and implementing the MentorAda program was not having enough budget. Therefore, these activities were developed with the resources available locally by each university, which in some cases were very limited. To minimize this situation, Uninorte sought strategic partners within the institution to support the program's implementation. Also, the savings obtained by implementing online activities made it possible to reduce the effects of the reduced budget. However, despite having mitigated this limitation, the financial resources play a key role in sustainability.

6.6 Conclusion

One of the most significant achievements of the W-STEM project at the consortium level is the integration achieved between universities in two world regions. Specifically, Europe and LA are dissimilar in some aspects but share, in some countries, situations of inequality in terms of the gender gap in STEM careers. For Uninorte, the methodology developed has allowed the achievement of the goals established in each phase of the project; the main results are the following:

The project promoted reflection about policies, mechanisms, and strategies in attracting, accessing, and advising women in STEM programs. This way, the institution can improve and adapt them to the specific circumstances of the Caribbean region.

Forming the W-STEM student group contributes to disseminating strategies for attracting young people to STEM programs and offers a space for support and counseling to those interested in these fields; this good practice will be replicated in other consortium members.

The team developed activities, tools, and support materials to foster the interest and motivation of high-school girls in STEM careers.

To support the orientation process of women in the first year of higher education in STEM programs, the team formed a group of mentors with female teachers and students in their final year of STEM programs.

The project has obtained comprehensive visibility, interest, and importance within Uninorte and in the Colombian Caribbean Region due to the activities developed for the three processes.

The activities and programs developed (attraction and recruitment campaigns, dissemination, scholarships) and in development (mentoring) and the currently active student group contribute to the project's sustainability once the ERASMUS + program funding ends. Moreover, the Uninorte team expects that the W-STEM project will contribute to reducing the gender gap in STEM careers in the Colombian Caribbean region.

Acknowledgements With the support of the Erasmus+ Programme of the European Union in its Key Action 2 "Capacity-building in Higher Education". Project W-STEM "Building the future of Latin America: engaging women into STEM" (Reference number 598923-EPP-1-2018-1-ES-EPPKA2-CBHE-JP). The content of this publication does not reflect the official opinion of the European Union. Responsibility for the information and views expressed in the publication lies entirely with the authors.

References

Aldrich, M., & Hall, P. (1980). American Association for the Advancement of Science. Office of Opportunities in Science (1980) Programs in Science, Mathematics, and Engineering for Women in the United States: 1966–1978.

American Society of Civil Engineers—ASCE. (2021). Diversity, Equity & Inclusion. https://www.asce.org/topics/diversity-equity-inclusion. Accessed 29 Nov 2021.

Bekki, J. M., Smith, M. L., Bernstein, B. L., & Harrison, C. (2013). Effects of an online personal resilience training program for women in stem doctoral programs. *JWM*, 19. https://doi.org/10.1615/JWomenMinorScienEng.2013005351.

Cantillo, V., & García, L. (2014). Gender and other factors influencing the outcome of a test to assess quality of education in civil engineering in Colombia. *Journal of Professional Issues in Engineering Education and Practice, 140*, 04013012. https://doi.org/10.1061/(ASCE)EI.1943-5541.0000194.

Carlucci Aiello, L. (2016). The multifaceted impact of Ada Lovelace in the digital age. *Artificial Intelligence, 235*, 58–62. https://doi.org/10.1016/j.artint.2016.02.003.

Ceci, S. J., Ginther, D. K., Kahn, S., & Williams, W. M. (2014). Women in academic science: A changing landscape. *Psychological Science in the Public Interest, 15*, 75–141.

Charlesworth, T. E., & Banaji, M. R. (2019). Gender in science, technology, engineering, and mathematics: Issues, causes, solutions. *Journal of Neuroscience, 39*, 7228–7243.

Crowley, M. F. (1977). *Women and minorities in science and engineering*. National Science Foundation.

Diekman, A. B., Brown, E. R., Johnston, A. M., & Clark, E. K. (2010). Seeking congruity between goals and roles: A new look at why women opt out of science, technology, engineering, and mathematics careers. *Psychological Science, 21*, 1051–1057. https://doi.org/10.1177/095679761 0377342.

Diekman, A. B., Clark, E. K., Johnston, A. M., et al. (2011). Malleability in communal goals and beliefs influences attraction to stem careers: Evidence for a goal congruity perspective. *Journal of Personality and Social Psychology, 101*, 902–918. https://doi.org/10.1037/a0025199.

Diekman, A. B., Weisgram, E. S., & Belanger, A. L. (2015). New routes to recruiting and retaining women in STEM: Policy implications of a communal goal congruity perspective. *Social Issues and Policy Review, 9*, 52–88.

Eagly, A. H. (1987). John M. MacEachran memorial lecture series; 1985. Sex differences in social behavior: A social-role interpretation.

ECLAC, United Nations. (2014). Compacts for equality: towards a sustainable future. United Nations, Santiago, Chile.

Else-Quest, N. M., Hyde, J. S., & Linn, M. C. (2010). Cross-national patterns of gender differences in mathematics: A meta-analysis. *Psychological Bulletin, 136*, 103.

Ernesto, F. P., Anathea, B. L., Alessandro, B., & Kim, D. (2018). Measuring gender equality in science and engineering: the SAGA survey of drivers and barriers to careers in science and engineering. UNESCO Publishing.

Farrell, S., Guerra, R. C., Longo, A., & Tsanov, R. (2018). A Virtual community of practice to promote LGBTQ inclusion in STEM: Member Perceptions and Community Outcomes. In: *ASEE Annual Conference*.

Feingold, A. (1988). Cognitive gender differences are disappearing. *American Psychologist, 43*, 95.

Ferriman, K., Lubinski, D., & Benbow, C. P. (2009). Work preferences, life values, and personal views of top math/science graduate students and the profoundly gifted: Developmental changes and gender differences during emerging adulthood and parenthood. *Journal of Personality and Social Psychology, 97*, 517.

Foundation, N. S. (2008). Broadening participation at the National Science Foundation: A framework for action.

Funk, C., & Parker, K. (2018). Women and men in STEM often at odds over workplace equity.

Furst-Holloway, S., & Miner, K. (2019). ADVANCEing women faculty in STEM: Empirical findings and practical recommendations from National Science Foundation ADVANCE institutions. *Equality, Diversity and Inclusion: An International Journal, 38*, 122–130. https://doi.org/10.1108/EDI-03-2019-295.

García Peñalvo, F. J., Bello, A., Domínguez, Á., & Romero Chacón, R. M. (2019). *Gender Balance Actions* (pp. 31–41). Results from a World Café Conversation. Education in the knowledge society.

García-Holgado A, Mena J, García-Peñalvo, F. J. et al. (2020). Gender equality in STEM programs: a proposal to analyse the situation of a university about the gender gap. In: *2020 IEEE Global Engineering Education Conference (EDUCON)*. IEEE, pp. 1824–1830.

García-Ramos, L., Peña-Baena, R., & García-Holgado, A. et al. (2021). Empowering young women in the caribbean region in stem. In: *2021 IEEE Global Engineering Education Conference (EDUCON)*. IEEE, pp. 1087–1092.

Geary, D. C. (1996). Sexual selection and sex differences in mathematical abilities. *Behavioral and Brain Sciences, 19*, 229–247.

General Directorate of Communication Networks, Content and Technology of the European Commission. (2013). Women active in the ICT sector—Publications Office of the EU. https://op.europa.eu/en/publication-detail/-/publication/9153e169-bd6e-4cf4-8638-79e2e9 82b0a3. Accessed 29 Nov 2021.

Goldman, R. D., & Hewitt, B. N. (1976). The Scholastic Aptitude Test" explains" why college men major in science more often than college women. *Journal of Counseling Psychology, 23*, 50.

González-Pérez, S., Mateos de Cabo, R., & Sáinz, M. (2020). Girls in STEM: Is it a female role-model thing? *Frontiers in Psychology, 11*, 2204.

Gray, H., Lyth, A., McKenna, C., et al. (2019). Sex differences in variability across nations in reading, mathematics and science: A meta-analytic extension of Baye and Monseur (2016). *Large-Scale Assess Educ, 7*, 2. https://doi.org/10.1186/s40536-019-0070-9.

Halpern, D. F. (1989). The disappearance of cognitive gender differences: What you see depends on where you look. *American Psychologist, 44*, 1156–1158. https://doi.org/10.1037/0003-066X.44.8.1156.

Harris PL-S. (2020). Creating Interest in STEM for African American Girls by Implementing a STEM Preschool Program. PhD Thesis, Northcentral University.

Herrmann, S. D., Adelman, R. M., Bodford, J. E., et al. (2016). The effects of a female role model on academic performance and persistence of women in STEM courses. *Basic and Applied Social Psychology, 38*, 258–268.

Hyde, J. S. (2014). Gender similarities and differences. *Annual Review of Psychology, 65*, 373–398. https://doi.org/10.1146/annurev-psych-010213-115057.

Hyde, J. S. (2016). Sex and cognition: Gender and cognitive functions. *Current Opinion in Neurobiology, 38*, 53–56. https://doi.org/10.1016/j.conb.2016.02.007.

Hyde, J. S., & Mertz, J. E. (2009). Gender, culture, and mathematics performance. *Proceedings of the National Academy of Sciences, 106*, 8801–8807. https://doi.org/10.1073/pnas.0901265106.

Instituto de Estadística de la UNESCO. (2013). Clasificación Internacional Normalizada de la Educación (CINE) 2011, Revisión 2. Instituto de Estadística de la UNESCO.

Kang, H., Calabrese Barton, A., Tan, E., et al. (2019). How do middle school girls of color develop STEM identities? Middle school girls' participation in science activities and identification with STEM careers. *Science Education, 103*, 418–439.

Lindberg, S. M., Hyde, J. S., Petersen, J. L., & Linn, M. C. (2010). New trends in gender and mathematics performance: A meta-analysis. *Psychological Bulletin, 136*, 1123–1135. https://doi.org/10.1037/a0021276.

My Fair Physicist? Feminine Math and Science Role Models Demotivate Young Girls - Diana E. Betz, Denise Sekaquaptewa, 2012. https://doi.org/10.1177/1948550612440735. Accessed 29 Nov 2021.

Matthews, K., Why is Diversity in Engineering a Major Opportunity? - ASME. https://www.asme.org/topics-resources/content/why-is-diversity-in-engineering-a-major-opportunity. Accessed 30 Nov 2021.

Milgram, D. (2011). How to recruit women and girls to the science, technology, engineering, and math (STEM) classroom. *Technology and Engineering Teacher, 71*, 4.

Morton, T. R., & Parsons, E. C. (2018). # BlackGirlMagic: The identity conceptualization of Black women in undergraduate STEM education. *Science Education, 102*, 1363–1393.

Mosatche, H. S., Matloff-Nieves, S., Kekelis, L., & Lawner, E. K. (2013). Effective STEM programs for adolescent girls: Three approaches and many lessons learned. *Afterschool Matters, 17*, 17–25.

Olsson, M., & Martiny, S. E. (2018). Does exposure to counterstereotypical role models influence girls' and women's gender stereotypes and career choices? a review of social psychological research. *Frontiers in Psychology, 9*, 2264. https://doi.org/10.3389/fpsyg.2018.02264.

Pahnke, J., O'Donnell, C., & Bascopé, M. (2019). Using science to do social good: STEM education for sustainable development. Position paper developed in preparation for the second "International Dialogue on STEM Education" (IDoS). Proceedings of the Second "International Dialogue on STEM Education" (IDoS), Berlin, Germany, pp. 5–6.

Penner, A. M., & Paret, M. (2008). Gender differences in mathematics achievement: Exploring the early grades and the extremes. *Social Science Research, 37*, 239–253. https://doi.org/10.1016/j.ssresearch.2007.06.012.

Prieto-Rodriguez, E., Sincock, K., & Blackmore, K. (2020). STEM initiatives matter: Results from a systematic review of secondary school interventions for girls. *International Journal of Science Education, 42*, 1144–1161. https://doi.org/10.1080/09500693.2020.1749909.

Quigley, C. F., Herro, D., & Jamil, F. M. (2017). Developing a conceptual model of STEAM teaching practices. *School Science and Mathematics, 117*, 1–12. https://doi.org/10.1111/ssm.12201.

Rossi, A. S. (1965). Women in science: Why so few? *Science, 148*, 1196–1202. https://doi.org/10.1126/science.148.3674.1196.

Rossiter, M. W. (1993). The matthew matilda effect in science. *Social Studies of Science, 23*, 325–341. https://doi.org/10.1177/030631293023002004.

Smith, J. L., Handley, I. M., Rushing, S., et al. (2018). Added benefits: How supporting women faculty in STEM improves everyone's job satisfaction. *Journal of Diversity in Higher Education, 11*, 502–517. https://doi.org/10.1037/dhe0000066.

Stevens, S., Andrade, R., & Page, M. (2016). Motivating young native American students to pursue STEM learning through a culturally relevant science program. *Journal of Science Education and Technology, 25*, 947–960. https://doi.org/10.1007/s10956-016-9629-1.

Su, R., Rounds, J., & Armstrong, P. I. (2009). Men and things, women and people: A meta-analysis of sex differences in interests. *Psychological Bulletin, 135*, 859–884. https://doi.org/10.1037/a0017364.

Swafford, M., & Anderson, R. (2020). Addressing the gender gap: Women's perceived barriers to pursuing STEM careers. *Journal of Research in Technical Careers, 4*, 61–74.

Syed, M., & Chemers, M. (2011). Ethnic minorities and women in STEM: Casting a wide net to address a persistent social problem. *Journal of Social Issues, 67*, 435–441. https://doi.org/10.1111/j.1540-4560.2011.01708.x.

The State of U.S. Science and Engineering 2020 I NSF - National Science Foundation. https://ncses.nsf.gov/pubs/nsb20201/. Accessed 29 Nov 2021.

Weisgram, E. S., Dinella, L. M., & Fulcher, M. (2011). The role of masculinity/femininity, values, and occupational value affordances in shaping young men's and women's occupational choices. *Sex Roles, 65*, 243–258. https://doi.org/10.1007/s11199-011-9998-0.

World Economic Forum. (2016). The Future of Jobs: Employment, Skills and Workforce Strategy for the Fourth Industrial Revolution.

W-STEM Consortium. (2019a). W-STEM Self-assessment Matrix. https://doi.org/10.5281/ZENODO.3594822.

W-STEM Consortium. (2019b). W-STEM Process Mapping Template. https://doi.org/10.5281/ZENODO.3594845.

W-STEM Consortium. (2019c). W-STEM Benchmarking Questionnaire Form. https://doi.org/10.5281/ZENODO.3594858.

Yang, J. A., Sherard, M. K., Julien, C., & Borrego, M. (2021). Resistance and community-building in LGBTQ+ engineering students. *JWM, 27*. https://doi.org/10.1615/JWomenMinorScienEng.2021035089.

Zell, E., Krizan, Z., & Teeter, S. R. (2015). Evaluating gender similarities and differences using metasynthesis. *American Psychologist, 70*, 10–20. https://doi.org/10.1037/a0038208.

Zilberman, A., & Ice, L. (2021). Why computer occupations are behind strong STEM employment growth in the 2019–29 decade. *Computer*, 4.

Chapter 7
Women Retention in STEM Higher Education: Systematic Mapping of Gender Issues

Esmeralda Campos, Claudia Lizette Garay-Rondero, Patricia Caratozzolo, Angeles Dominguez, and Genaro Zavala

Abstract Science, Technology, Engineering, and Mathematics (STEM) in Higher Education (HE) helps foster students' motivation to continue studying and cultivates students' regard for the role of science and technology in society. The gender gap in STEM HE can reduce through institutional efforts; however, the underrepresentation of women is prevalent. There have been efforts to research and implement strategies to increase the number of people attending STEM fields with a specific action to attract and retain women in these areas. Hence, the purpose of this research work is to carry out and show the results of a Systematic Mapping (SM) related to how HE institutions aim to address the gender gap in STEM education through research and educational innovation. The SM focused on published work from 2011 to 2021 indexed in Web of Science or Scopus. Findings show the state of knowledge for an essential topic: reducing the gender gap through guidance and retention strategies to attain completion. Furthermore, descriptive results give a general overview of the area, relevant trends, and other analytical evidence that provides an in-depth understanding of HE institutions' needs. We conclude that the retention of women studying STEM HE

E. Campos
Institute for the Future of Education, Tecnologico de Monterrey, Monterrey, Mexico
e-mail: esmeralda.campos@tec.mx

C. L. Garay-Rondero · P. Caratozzolo · G. Zavala
Institute for the Future of Education, School of Engineering and Sciences, Tecnologico de Monterrey, Monterrey, Mexico
e-mail: clgaray@tec.mx

P. Caratozzolo
e-mail: pcaratozzolo@tec.mx

G. Zavala
e-mail: genaro.zavala@tec.mx

A. Dominguez (✉)
Institute for the Future of Education, School of Medicine and Health Sciences, Tecnologico de Monterrey, Monterrey, Monterrey, Mexico
e-mail: angeles.dominguez@tec.mx

A. Dominguez · G. Zavala
School of Engineering, Universidad Andres Bello, Santiago, Chile

© The Author(s) 2022
F. J. García-Peñalvo et al. (eds.), *Women in STEM in Higher Education*, Lecture Notes in Educational Technology, https://doi.org/10.1007/978-981-19-1552-9_7

has become an essential issue worldwide universities have addressed increasingly during the last decade.

Keywords Retention · STEM education · Gender perspective · Guidance · Higher education · Educational innovation

7.1 Introduction

Science, Technology, Engineering, and Mathematics (STEM) in Higher Education (HE) has many objectives such as promoting learning, fostering students' motivation to continue studying, and cultivating students' regard for the role of science and technology in society. Research has proven that there is a gender gap in STEM HE due to several factors such as the traditional role of women in society and the lack of female role models and mentors in STEM areas, to name a few (Wang & Degol, 2013). Several international organizations have emphasized the importance of addressing the gender gap in STEM HE explicitly. Addressing the gender gap in HE is part of the UN's Sustainable Development Goal 4, specifically for target 4.3, which refers to ensuring equal access for women and men to tertiary education, including university. According to the Organization for Economic Co-operation and Development (OECD), the gender gap in STEM HE enrollment differs for different fields of study; the field of natural sciences, mathematics, and statistics has reached gender parity, while for engineering and information and communication technologies (ICT) the gender gap persists (Organisation for Economic Co-operation and Development, 2021). It is noteworthy that among the mentioned fields, there may exist differences between countries and areas of study, such as biology and physics within the natural sciences (García-Peñalvo, 2019). The OECD highlights the importance of removing stereotypes and implementing policies for reducing the gender gap in different fields of study.

HE institutions can implement several strategies to address the gender gap in STEM areas through different processes: attraction, access, and retention (García-Peñalvo et al., 2019). Within the retention process, some universities may identify guidance, completion, and other similar terms. Retention of engineering students is an important issue that academic institutions must address to avoid dropout by creating an inclusive and supportive environment (García-Holgado et al., 2020). We have identified review articles in the literature about the retention and guidance of women in STEM careers in HE. These review articles have provided some insights into the factors and obstacles that women in STEM face throughout their careers and their persistence in pursuing STEM majors. Eddy and Brownell (Eddy & Brownell, 2016) contributed to the literature with a framework that identifies performance and engagement as observable inequalities to explain gender gaps on persistence in STEM, a measure correlated with retention (Eddy & Brownell, 2016). Women in STEM face challenges in predominantly masculine environments, such as gendered organizational culture and stereotypes; their coping strategies center on conforming, impression management, and proactivity (Makarem & Wang, 2020). Within the STEM careers, some authors identify Physics, Engineering, Mathematics, and Computer science (PEMC) as more related to mathematical ability than the biological sciences fields. Within this distinction, a study reported that girls' mathematical ability beliefs under challenges may impact their chance of majoring in PEMC fields (Perez-Felkner et al., 2017). A study on the NSF ADVANCE program identified that systemic obstacles to recruitment, retention, and promotion impact women's underrepresentation in STEM areas (DeAro et al., 2019). In general, these articles confirm that the gender gap in STEM education is a systemic issue and that academic institutions can implement strategies in their guidance and retention process to avoid women's dropout from STEM careers.

Academic institutions and policymakers must inform each other about the strategies to guide women in STEM careers and avoid academic dropout (García-Peñalvo et al., 2019). We identify the need to conduct a literature review focused on the retention strategies that universities worldwide have adopted to guide and retain women in their STEM programs. This study aims to provide a systematic mapping of the literature related to how HE institutions aim to address the gender gap in STEM education through research and educational innovation. The main contribution of this work to society is to inform about best practices for women's retention in STEM careers. Authors analyze the state of knowledge in the last ten years concerning geographic information, relevant authors, and trending research topics with this aim in mind. In the methodology, we present the PRISMA method that we follow to develop the literature mapping. Then, the Results and Discussion section provides database analysis and the discussion of the emerging topics. The final section concludes with specific recommendations for HE institutions and policymakers to address the gender gap in STEM HE through the guidance and retention process.

7.2 Methodology

The literature search should be designed to be robust and reproducible to ensure the minimization of biases. Although there are alternative approaches for both conducting literature searches and reporting systematic reviews or mappings, the authors have encountered difficulties with some of those methodologies due to the following reasons:

(i)   the various methods are not comparable as they share few common reporting elements

(ii) there are numerous new checklists and tools that are not sufficiently described to be reliably replicated

(iii) there is a debate about what constitutes a reproducible search and how best to report the details of the search

The use of non-validated bibliographic search methods can raise doubts and reduce confidence in the final conclusions of the systematic review or mapping (Moher et al., 2015). The authors noted that if readers cannot understand or reproduce how the information was collected for this systematic mapping, they may suspect that the authors have introduced bias by not conducting a comprehensive or pre-specified literature search. These were the reasons why the authors chose the most commonly used reporting guide for systematic reviews and mappings, which covers the literature search component, which is the Preferred Reporting Items Statement for Systematic Reviews and Meta-Analyses, known as PRISMA Statement. The methodological approach consisted of a systematic mapping of the literature with a review process based on the PRISMA protocol (Xiao & Watson, 2019). The sequence of the PRISMA process consisted of 27 items (Page et al., 2021). The methodological stage of PRISMA is made up of items from 5 to 15, which are the following: Eligibility criteria; Information sources; Search strategy; Selection process; Data collection process; Data items; Study risk of bias assessment; Effect measures; Synthesis methods; Reporting bias assessment; Certainty assessment. In the present study, these items were appropriately included in the following steps:

1. Formulation of the problem and definition of the Research Questions (RQ).
2. Developing a review protocol.
3. Systematic search of literature, including the following steps: select databases and their descriptors; derive keywords from research questions; and adopt a sampling logic.
4. Screening for inclusion and exclusion criteria.
5. Extracting, analyzing, and synthesizing data.

The literature mapping carried out in the present study was guided by three research questions (RQ). From them, the methodology for data extraction was designed and the report of the findings was written. The RQs were narrowed from a general research topic until we chose a more defined subtopic from the original search using mapping (Campos et al., 2020). In this way, it was possible to identify the activities involving the refined RQs. The chapter structure and its representation of the data and findings are determined from the following RQs to bring out the information searching to achieve the main objective of this study: To carry out and show the results of a Systematic Mapping (SM) related to how HE institutions aim to address the gender gap in STEM education through research and educational innovation. Each RQ is defined to reach the specific Research Objectives (ROs) in this work as given below.

Descriptive research objectives and research questions are as follows:

- *RO1:* Identify the related research works disseminated between 2011 and 2021 and determine how they have been distributed in the defined databases.

  RQ1: How are publications distributed in the defined databases? What is the distribution of publications in the period between 2011 and 2021?

- *RO2*: Identify the grouping of the main research works in terms of the document type.

RQ2: What is the distribution by type of document?

- *RO3*: Identify the journals and conferences with distinguished publications related to this topic.

RQ3: Which are the journals with the largest publications on this topic? Which conferences contribute the most to the literature on this topic?

- *RO4*: Determine the geographical distribution of the first authors and which are the specific countries that have been contributing to this research area.

RQ4: What is the geographical distribution of the first authors? Which are the countries that carry out research on the research topic?-

Analytical research objectives and research questions:

- *RO5*: Recognize the significant trending words stated in the obtained works and their accumulated frequency displayed in the keywords. Consequently, determine how these keywords can inform the trends in women retention in STEM HE.

RQ5: What are the trending words (most common keywords) by the authors? What is the frequency of our keywords in the abstract, and how can these keywords inform the trends in publications on women retention in STEM HE?

- *RO6*: Characterize the most relevant articles by their citation metrics, showing the most significant impact on the research topic.

RQ6: What are the most cited articles? Which publications have had the greatest impact in the area?

Due to the importance of selecting a review protocol, we decided to choose the PRISMA protocol which allowed us to reduce the possibility of research bias in data selection and subsequent extraction and analysis (Moher et al., 2015). In order to validate the protocol before executing the search, the work team was divided into two teams that worked independently to perform a peer review and validate the sequence of the protocol, as shown in Fig. 7.1.

The keywords for the search were derived from the RQs. For the first step in the literature search, the keywords "women in STEM", "STEM education" and "STEM learning" were used. Some concepts used as keywords included synonyms, alternative spellings, and related words, as in the case of gender gap, retention, guidance, dropout, completion, and attrition. The first search query for both databases was formed by the following general terms (linked by AND operators) and synonyms (linked by OR operators):

A.   STEM Education

   A1. STEM learning

   A2. Women in STEM

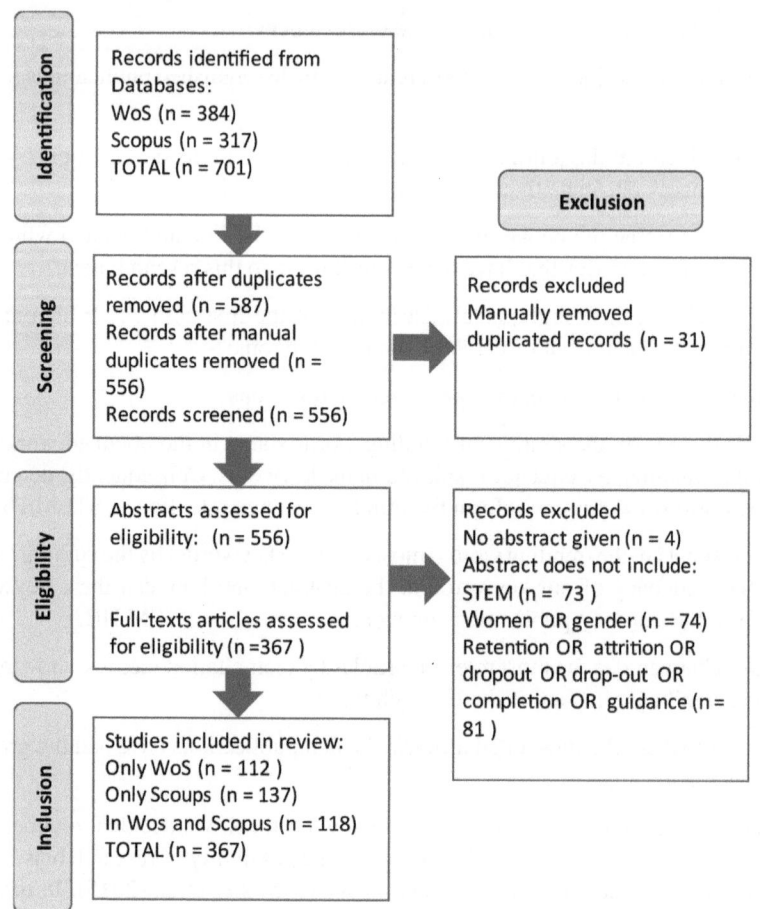

Fig. 7.1 PRISMA 2020 flow diagram for literature search

B. Gender Gap

    B1. Women

    B2. Gender

C. Higher Education

    C1. University

    C2. College

D. Retention

    D1. Guidance

    D2. Dropout

D3. Completion

D4. Attrition

Using these terms and synonyms, the query has the general form:

(A OR A1 OR A2) AND (B OR B1 OR B2) AND (C OR C1 OR C2) AND (D OR D1 OR D2 OR D3 OR D4).

To determine the relevance of each manuscript, the titles were analyzed and if the content discussed the topic of women in STEM or gender gap in STEM, we obtained their full reference, including author, year, title, and abstract, to then complete future evaluations. It is important to highlight that, in this first stage, all those manuscripts that were related to the botanical meanings "cell" of the word "stem" were detected and excluded. The search was limited to two databases: Web of Science and Scopus, as they are the two main databases frequented by researchers from the engineering and science disciplines (Pranckutė, 2021).

Due to the changes in the perception and revaluation of the role of women in STEM areas and the impulse that the issue of the gender gap has had in society in general, we limit the publication date to 2011 and 2021 (articles published in the last ten years), so that we can build our review on the recent literature that considers the influence of political, social, and cultural movements on the inclusion of women in STEM. The inclusion and exclusion criteria are defined in the PRISMA protocol in Fig. 7.1.

Following the process detailed below, we refined the search queries used in each database. To allow for replicability, we include the exact final search queries used in Scopus and Web of Science in the Appendix. In the Scopus database, the query yielded 255 results. In the WoS database, the final query yielded 230 results. The final query has the following general form, where the general terms are linked by AND operators and the synonyms are linked with OR operators:

Title, Abstract or keywords include:

A.   STEM Education

   A1. STEM learning

   A2. Women in STEM

B.   Gender Gap

   B1. Women

   B2. Gender

C.   Higher Education

   C1. University

   C2. College

D.   Retention

   D1. Guidance

D2. Dropout

D3. Drop out

D4. Completion

D5. Attrition

Abstract includes:

E.   STEM
F.   Gender

F1. Women

G.   Retention

G1. Attrition

G2. Guidance

G3. Dropout

G4. Drop out

G5. Completion

Time filter: 2011 to 2021.
Subject area filters.
Using these terms and synonyms, the final query has the general form:
Title, Abstract, Keywords ((A OR A1 OR A2) AND (B OR B1 OR B2) AND (C OR C1 OR C2) AND (D OR D1 OR D2 OR D3 OR D4 OR D5)).
AND Abstract (E AND (F OR F1) AND (G OR G1 OR G2 OR G3 OR G4 OR G5)).
AND Time (2011 to 2021).
AND Subject area filters.

7.3   Results and Discussion

Regarding the PRISMA stage corresponding to results, this study offers preliminary results corresponding to the study selection items (Tricco et al., 2018), which correspond, respectively, to:

16a. Describe the results of the search and selection process, from the number of records identified in the search to the number of studies included in the review, ideally using a flow diagram.

16b. Cite studies that might appear to meet the inclusion criteria, but which were excluded, and explain why they were excluded.

For reasons concerning the length of this chapter, items 17 to 22 (from the Results Section of the PRISMA Checklist) were excluded from this report. The items not considered were: Study characteristics; Risk of bias in studies; Results of individual studies; Results of syntheses; Reporting biases; and, Certainty of evidence.

We present the results of our analysis of the 367 records based on the databases, the type of documents with an overview of the journals and conferences that publish the most about this topic, the countries of the first authors that publish research on this topic, and the emerging keywords to find interesting trends.

7.3.1   Database Analysis and Timeline

We analyzed the articles that emerged in the Scopus database, in the Web of Science, and in both. We found that 118 records were in both databases, so 112 records were exclusively in WoS and 137 records were exclusively in Scopus, coherent with Fig. 7.1. We found an increasing trend in the number of records published between 2011 and 2021. From 2011 to 2013, there were less than 20 records published each year. From 2014 to 2017, there was a turning point in the number of records, having between 20 and 35 each year. After 2018, the number of records increased to more than 50 each year until 2020. In 2021, we found 32 records up until the middle of the year when we created this database, so the trend is confirmed. In each database, we see similar trends. In the Scopus database, the years with most records were 2017 with 28, 2018 with 41, and 2020 with 45 records. In Web of Science, the most relevant years were 2018 with 29, 2020 with 31, and 2019 with 41 records. These results are promising because they show that the research about guidance and retention of women in STEM HE is increasing each year. This implies that universities are more interested in implementing strategies for reducing the gender gap in STEM HE completion than 10 years ago.

7.3.2   Type of Documents

We analyzed the types of documents published about this topic in Scopus and WoS between 2011 and 2021. We identified 196 articles (53%), 159 conference papers (43%), 10 book chapters (3%), and 2 books (1%). We present an analysis of the publishing trends for articles and conference papers based on the years and the most relevant journals and conferences on the topic.

We can see an increasing trend in the number of articles in Fig. 7.2; from 2011 to 2014 and 2016, there were less than 10 articles per year. In 2015 and 2017, there were more than 15 articles, in 2018, there were more than 20, and from 2019 to 2021 (ongoing), there were more than 30 articles each year. The journals that have published more on this topic were PLOS ONE (9 articles), the International Journal of STEM Education (7 articles), Equality, Diversity and Inclusion, Frontiers in Psychology, Journal of Chemical Education, Journal of Women and Minorities in Science and Engineering, Sex Roles, and Teachers College Record (5 articles, each).

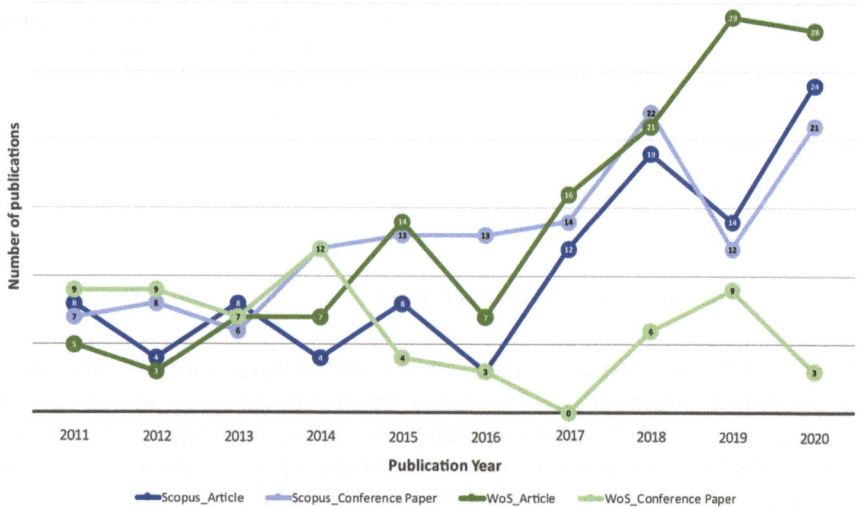

Fig. 7.2 Publication trends from 2011 to 2020 per type of document (journal article or conference paper) and source (Scopus or WoS). *Note* The graph only includes full years (2011 to 2020). That is, 2021 is excluded because the information is not completed yet.

The conference papers have a steadier trend, remaining between 10 and 20 conference papers every year, except for 2018 and 2020, where we found 26 and 22, respectively. The conference that contributes the most to this area is the American Society for Engineering Education (ASEE) Annual Conference and Exposition with 99 conference papers (27% of all records), followed by the Frontiers in Education (FIE) Conference, with 13 papers (3.5% of all records). Other relevant conferences were the International Conference on Education and New Learning Technologies (EDULEARN), the International Conference of Education, Research and Innovation (ICERI), and the International Technology Education and Development (INTED) Conference, contributing 4 conference papers each.

7.3.3 Geographic Data

We analyzed the country of the first author for each article based on their affiliation. We found that the vast majority of studies about retention of women in STEM programs are performed in the United States of America (295 articles, 80.7%). We found other 6 countries that contribute to the literature on this topic: Australia (9 articles, 2.5%), Canada (8 articles, 2.2%), Germany (7 articles, 1.9%), Mexico (5 articles, 1.4%), Spain (4 articles, 1.1%), and the United Kingdom (4 articles, 1.1%). We identified 24 more countries that have contributed to the literature with less than 3 articles each, representing less than 1% of the total. As the data suggests, most of the research on this topic is done in North America, specifically in the US, Canada, and Mexico, followed by Europe with Germany, Spain, and the UK. Most of these countries are English or Spanish speaking.

7.3.4   Trending Keywords

In an effort to capture the authors' most common keywords as an indicator of a trend, we analyzed the keywords given by the authors of the 367 documents. We present them in Fig. 7.3. STEM (99 times) and gender (80 times) are the words that appear most frequently (8.2% and 6.6%, respectively). Then, there is a step down to 3.6% (44 times) for the word education and 3.3% (40 times) for women. The next group of most common keywords is science (25 times, 2.1%), diversity (24 times, 2.0%), and engineering (23 times, 1.9%). Then comes faculty, students' retention, academic, and

Fig. 7.3 Analysis of the authors' most common keywords

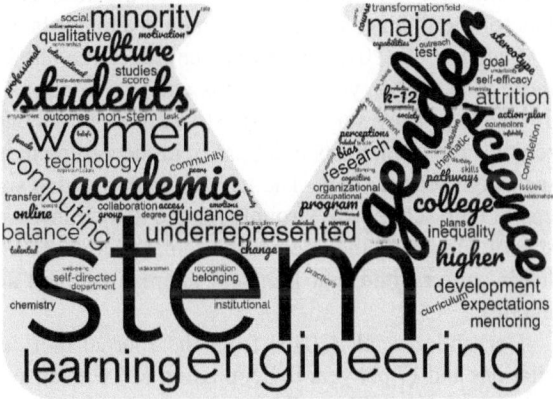

learning with about 1.8% each (around 18 mentions each of these keywords). The last group consists of the keywords career, culture, equity, gap, persistence, minority, major appearing 12 times (corresponding to 1.0% for each one).

The diversity and science keywords are of particular interest since they could come from papers with different perspectives. From the 24 papers, eight of them (33%) study women with a race perspective; they are studies in which they focus on minorities (e.g., Black American, Hispanics) and gender. Three manuscripts study the women's perspective on a broader diverse standpoint about race and GLBT or disabilities. The rest of them (54%) regard women from only the gender perspective. In the case of the science keyword, three (12%) of them focus on biology, four focus on computer science, and five focus on natural and exact sciences. The rest (52%) focus on STEM disciplines without specifying a precise science of emphasis.

7.3.5 Most Cited Articles

There are six documents that are cited by more than 100 publications. Ong et al. (Ong et al., 2011), a paper from the Harvard Educational Review in 2011 has 401 citations averaging more than 36 citations per year in Scopus and 292 citations in WoS with an average of more than 26 citations per year. They produced a nice review of research in the previous 40 years regarding the African American females' experiences in the STEM fields in their undergraduate and graduate education. They made a strong discussion of policy implications of their work and they emphasized what were the topics where research was needed. The number of citations has to do with that discussion.

Cech et al. (2011), from the American Sociological Review, has 287 citations averaging more than 28 citations per year in Scopus and 243 with an average of more than 24 citations per year in WoS. They made a critical contribution regarding the retention of women in STEM fields at the undergraduate level. In their data, instead of having family plans or math self-assessment as key factors, they introduced a new factor at that time, professional role confidence. They found that women's lack of this role confidence compared to men is a major factor for women attrition.

Sadler et al. (2012) from Science Education has 282 citations and an average of more than 28 per year in Scopus and 206 citations and an average of more than 18 in WoS. They studied 6000 students in 34 colleges around the USA to see the relationship of interests in STEM fields according to how that interest shifted during their high school years. The key factor they found was that females' interest at the end of high school was strongly related to their interest at the beginning of their high school. However, they compared females' and males' decline in interest in STEM fields and resulted that females' interest declined significantly more than males' interest.

Hernandez et al. (2013) from the Journal of Educational Psychology has 153 citations and 17 per year in Scopus and 125 with an average of more than 13 citations per year in WoS. They did a three-year study following university high academic achievers from the minority groups in the STEM fields. They found that students' engagement in undergraduate research was a key factor for success. Related to this, they found that the growth in scientific self-identity and task goals capabilities has a strong effect on students' retention and that performance-avoidance goals were related to students' attrition.

Morgan et al. (2013) from Social Science Research has 126 citations with an average of 14 per year in Scopus and a total of 114 with an average of more than 12 citations per year in WoS. They did a longitudinal study for four years focusing on students' pathways through college. They found gender differences across the STEM fields including health science disciplines. They also found that one of the main factors for students' college selection was students' occupational plans as seniors in high school.

Dennehy and Dasgupta (2017) is the most recent document in the list of the most cited with 121 citations averaging more than 24 per year in Scopus and 92 with an average of more than 18 citations per year in WoS. They did a longitudinal intervention study for two years in which, in the first year, entering female students participated in a mentor program. Some of the students had a female mentor, some of them a male mentor, and the rest had no mentor at all. They found that retention was more successful with a same-sex mentor and that result was modulated by an increase of belonging and self-efficacy.

Interestingly, from the six documents that had more citations, four of them were studies in which the authors were interested in finding factors in which retention depends on minority groups, in particular women. Two of them even took data of students when they were in high school. Another document, the one with more citations is a review analyzing more than 110 previous studies. We highlight that those five studies had no intervention. The last document, the most recent one is the only study with an intervention with significant results.

The descriptive and analytical trends presented in this results section have provided an overview of the state of knowledge regarding institutional efforts to guide and retain women in STEM fields. From our results, we can envision that women retention in STEM HE is an important issue that universities are beginning to address in different parts of the world. The analysis has shown the countries that have contributed the most, and the keywords that greatly define this area of research. The last part of the results has focused on an in-depth analysis of the most cited articles, finding interesting trends there, such as the focus on studies about women retention, but fewer articles about actual implementations.

7.4 Conclusion

Retention of women studying HE in the STEM areas has become an important issue addressed by many universities in different countries. Especially since 2016, there is a clear increasing trend in journal articles and conference papers dedicated to women or the gender gap within STEM education with a focus on retention, attrition, dropout, guidance or completion. This systematic literature mapping reveals how HE institutions show concern and awareness by documenting their good practices and institutional policies. This study analyzes the dropout issues in HE from a gender perspective within 2011 and half of 2021. It could be used as a reference source for further investigations or to expand the current one, since this revision focused on the publications reported in the two main indexing systems Web of Science and Scopus.

We conclude there is an increasing interest in promoting retention of women studying HE in the STEM areas by the number of articles each year. However, the interest of institutions and researchers in the field is a small number compared to the total number of universities that offer those programs. We recommend institutions adopt new policies addressing this problem from the action perspective; that is, getting to know the research results and implementing attraction, access, and retention actions to increase the number of women involved. In attraction, we recommend implementing a campaign directed to young women with seminars and talks by professional women in STEM areas. In access, we recommend giving close follow-ups to young women for college/university applications. In retention, we recommend offering guidance programs such as mentoring and academic consultation, fostering identity by creating women's networks, and offering seminars and workshops by women in the industry or outside the academic system.

This research provides a perspective on women retention studies in HE institutions within the STEM fields and their approaches to support their students to complete their academic programs. This chapter may be of interest to researchers, students, teachers, and decision makers interested in improving the academic environment and culture to promote equity for all.

Acknowledgements The authors would like to thank the Institute for the Future of Education at Tecnologico de Monterrey for its support through the Interdisciplinary Research Group named Socially Oriented Interdisciplinary STEM Education (SOI-STEM).

Appendix

Final query for Scopus Database:

((ABS(retention) OR ABS(attrition) OR ABS(guidance) OR ABS(dropout) OR ABS(drop out) OR ABS(completion)) AND (ABS(gender) OR ABS(women)) AND (ABS(STEM))) AND (TITLE-ABS-KEY((stem OR "STEM Education" OR "STEM learning" OR "Women in STEM" AND NOT cell) AND ("gender gap" OR women OR gender) AND ("higher education" OR university OR college) AND (retention OR guidance OR dropout OR "drop out" OR completion OR attrition)) AND (LIMIT-TO(PUBYEAR, 2021) OR LIMIT-TO(PUBYEAR, 2020) OR LIMIT-TO(PUBYEAR, 2019) OR LIMIT-TO(PUBYEAR, 2018) OR LIMIT-TO(PUBYEAR, 2017) OR LIMIT-TO(PUBYEAR, 2016) OR LIMIT-TO(PUBYEAR, 2015) OR LIMIT-TO(PUBYEAR, 2014) OR LIMIT-TO(PUBYEAR, 2013) OR LIMIT-TO(PUBYEAR, 2012) OR LIMIT-TO(PUBYEAR, 2011)))

Final query for Web of Science Database:

(AB=(stem) AND (AB=(retention) OR AB=(attrition) OR AB=(guidance) OR AB=(dropout) OR AB=(drop out) OR AB=(completion)) AND (AB=(women) OR AB=(gender))) AND (((((ALL=("stem education") OR ALL=("stem learning") OR ALL=(women in stem)) AND (ALL=("gender gap") OR ALL=(women) OR ALL=(gender))) NOT ALL=(CELL)) AND (ALL=("HIGHER EDUCATION") OR ALL=(UNIVERSITY) OR ALL=(COLLEGE)) AND (ALL=(RETENTION) OR ALL=(GUIDANCE) OR ALL=(DROPOUT) OR ALL=("DROP OUT") OR ALL=(COMPLETION) OR ALL=(ATTRITION))) AND ((PY==("2021" OR "2020" OR "2019" OR "2018" OR "2017" OR "2016" OR "2015" OR "2014" OR "2013" OR "2012" OR "2011")) NOT (DT==("CORRECTION" OR "MEETING ABSTRACT") OR TASCA==("ORTHOPEDICS" OR "OBSTETRICS GYNECOLOGY" OR "MEDICINE GENERAL INTERNAL" OR "CLINICAL NEUROLOGY" OR "PSYCHIATRY" OR "REHABILITATION" OR "HEALTH CARE SCIENCES SERVICES" OR "SPORT SCIENCES" OR "EVOLUTIONARY BIOLOGY" OR "SURGERY" OR "GERIATRICS GERONTOLOGY" OR "PUBLIC ENVIRONMENTAL OCCUPATIONAL HEALTH" OR "ONCOLOGY" OR "ENDOCRINOLOGY METABOLISM" OR "PLANT SCIENCES" OR "LANGUAGE LINGUISTICS" OR "LINGUISTICS" OR "RADIOLOGY NUCLEAR MEDICINE MEDICAL IMAGING" OR "NURSING" OR "SOCIAL WORK" OR "SUBSTANCE ABUSE" OR "GERONTOLOGY" OR "LITERATURE" OR "HOSPITALITY LEISURE SPORT TOURISM" OR "UROLOGY NEPHROLOGY" OR "TROPICAL MEDICINE" OR "CRIMINOLOGY PENOLOGY" OR "HEALTH POLICY SERVICES" OR "INTEGRATIVE COMPLEMENTARY MEDICINE"))))

References

Campos, E., Armenta, I. H., Barniol, P., & Ruiz, B. (2020). Physics education: Systematic mapping of educational innovation articles. *Journal of Turkish Science Education, 17*(3), 315–331.

Cech, E., Rubineau, B., Silbey, S., & Seron, C. (2011). Professional role confidence and gendered persistence in engineering. *American Sociological Review, 76*(5), 641–666. https://doi.org/10.1177/0003122411420815.

DeAro, J., Bird, S., & Ryan, S. M. (2019). NSF ADVANCE and gender equity: Past, present and future of systemic institutional transformation strategies. *Equality, Diversity and Inclusion: An International Journal, 38*(2), 131–139.

Dennehy, T. C., & Dasgupta, N. (2017). Female peer mentors early in college increase women's positive academic experiences and retention in engineering. In *Proceedings of the National Academy of Sciences of the United States of America*, vol. 114(23), 5964–5969. https://doi.org/10.1073/pnas.1613117114.

Eddy, S. L., & Brownell, S. E. (2016). Beneath the numbers: A review of gender disparities in undergraduate education across science, technology, engineering, and math disciplines. *Physical Review Physics Education Research, 12*(2), 020106.

García-Holgado, A., González-González, C. S., & Peixoto, A. (2020). A comparative study on the support in engineering courses: A case study in Brazil and Spain. *IEEE Access, 8*, 125179–125190.

García-Peñalvo, F. J., Bello, A., Dominguez, A., & Romero Chacón, R. M. (2019). Gender balance actions, policies and strategies for STEM: Results from a world café conversation. *Education in the Knowledge Society, 20*, 31–41.

García-Peñalvo, F. J. (2019). Women and STEM disciplines in Latin America: The W-STEM European Project. *Journal of Information Technology Research, 12*(4), v–viii.

Hernandez, P. R., Schultz, P. W., & Estrada, M. et al. (2013). Sustaining optimal motivation: A longitudinal analysis of interventions to broaden participation of underrepresented students in STEM. *Journal of Educational Psychology, 105*(1), 89–107. https://doi.org/10.1037/a0029691.

Makarem, Y., & Wang, J. (2020). Career experiences of women in science, technology, engineering, and mathematics fields: A systematic literature review. *Human Resource Development Quarterly, 31*(1), 91–111.

Moher, D., Shamseer, L., & Clarke, M. et al. (2015). Preferred reporting items for systematic review and meta-analysis protocols (PRISMA-P) 2015 statement. https://doi.org/10.1186/2046-4053-4-1.

Morgan, S. L., Gelbgiser, D., & Weeden, K. A. (2013). Feeding the pipeline: Gender, occupational plans, and college major selection. *Social Science Research, 42*(4), 989–1005. https://doi.org/10.1016/j.ssresearch.2013.03.008.

Ong, M., Wright, C., Espinosa, L. L., & Orfield, G. (2011). Inside the double bind: A Synthesis of empirical research on undergraduate and graduate women of color in science, technology, engineering, and mathematics. *Harvard Educational Review, 81*(2), 172–208. https://doi.org/10.17763/haer.81.2.t022245n7x4752v2

Organisation for Economic Co-operation and Development. (2021). Education at a Glance 2021: OECD Indicators. *OECD Publishing, Paris,*. https://doi.org/10.1787/b35a14e5-en

Page, M. J., McKenzie, J. E., & Bossuyt, P. M. et al. (2021). The PRISMA 2020 statement: an updated guideline for reporting systematic reviews. *BMJ, 372*.

Perez-Felkner, L., Nix, S., & Thomas, K. (2017). Gendered pathways: How mathematics ability beliefs shape secondary and postsecondary course and degree field choices. *Frontiers in Psychology, 8*, 386.

Pranckutė, R. (2021). Web of science (Wos) and scopus: The titans of bibliographic information in today's academic world. In Publications (Vol. 9, Issue 1). MDPI AG. https://doi.org/10.3390/publications9010012.

Sadler, P. M., Sonnert, G., Hazari, Z., & Tai, R. (2012). Stability and volatility of STEM career interest in high school: A gender study. *Science Education, 96*(3), 411–427. https://doi.org/10.1002/sce.21007.

Tricco, A. C., Lillie, E., Zarin, W., et al. (2018). PRISMA extension for scoping reviews (PRISMA-ScR): Checklist and explanation. *Annals of Internal Medicine, 169*(7), 467–473.

Wang, M. T., & Degol, J. (2013). Motivational pathways to STEM career choices: Using expectancy-value perspective to understand individual and gender differences in STEM fields. *Developmental Review, 33*(4), 304–340.

Xiao, Y., & Watson, M. (2019). Guidance on conducting a systematic literature review. *Journal of Planning Education and Research, 39*(1), 93–112. https://doi.org/10.1177/0739456X17723971

Chapter 8
Mentoring Female Students in Engineering as a Way of Caring

Patricia Jiménez, Jimena Pascual, and Andrés Mejía

Abstract In this chapter, we report on a mixed research study about the ways mentors attribute meaning and purpose to the practice of mentorship in a program for first and second-year female students of industrial engineering at Pontificia Universidad Católica de Valparaíso (Chile). A quantitative analysis (principal components) of 28 mentors' responses to a questionnaire about their motivations to be mentors prompted us to examine more deeply one of the factors found. Motivations constitutive of this factor referred to mentors' ways of attributing meaning and purpose to their practice, which can be understood in terms of caring. Four focus groups with 13 mentors in total allowed us to advance further into the examination of what the good pursued by this practice of mentorship meant for them. In our analysis, we drew from Tronto's four phases of care: caring about, caring for, caregiving and care receiving, and their corresponding defining moral elements: respectively, attentiveness, responsibility, competence, and responsiveness.

Keywords Women in STEM · Mentoring · Higher education · Ethics of care · Motivations to mentor · Retention

8.1 Introduction

The underrepresentation of women in Science, Technology, Engineering, and Mathematics (STEM) careers, especially in engineering, is well documented in the literature (Cadaret et al., 2017; Dennehy & Dasgupta, 2017; Hernandez et al., 2017; Koul, 2018). UNESCO (2017) reports that the global average share of female

P. Jiménez (✉) · J. Pascual
Pontificia Universidad Católica de Valparaíso, Valparaíso, Chile
e-mail: patricia.jimenez@pucv.cl

J. Pascual
e-mail: jimena.pascual@pucv.cl

A. Mejía
Universidad de Los Andes, Bogotá, Colombia
e-mail: jmejia@uniandes.edu.co

© The Author(s) 2022
F. J. García-Peñalvo et al. (eds.), *Women in STEM in Higher Education*, Lecture Notes in Educational Technology, https://doi.org/10.1007/978-981-19-1552-9_8

students enrolled in higher education in engineering, manufacturing, and construction programs is 27%. In Chile, the 2020 enrollment of women in technology-based careers (including engineering) is 24% (MCTCI, 2020).

This phenomenon is generally seen as a problem that goes beyond a feminist claim for equity. Indeed, the lack of diversity in the STEM workforce has been argued to have negative consequences for scientific innovation, creativity, and social relevance (Hernandez et al., 2017). Therefore, incorporating more women would help broaden the diversity of product design and problem-solving perspectives and add greater rigor in decision-making. In addition, this increase would help satisfy the global shortage of engineering professionals (Smith, 2017; Steenkamp et al., 2017; Stelter et al., 2021) and, because of the present feminization of poverty, would impact poverty reduction. Broadening the participation of women in STEM disciplines calls for new learning/teaching methodologies and new support programs that enhance college experiences and other motivators for academic persistence.

The factors that explain this lower participation are varied: gender/occupation stereotypes (Bonaldi & Silva, 2014; Cadaret et al., 2017; Powell et al., 2012; Salikutluk & Heyne, 2017; Stout et al., 2011), beliefs about competence and self-efficacy (Cadaret et al., 2017; Cech et al., 2011; Falco & Summers, 2019; Marra et al., 2009); previous academic preparation, prosocial orientation, school and family socialization from an early age, and lack of role models in engineering, among others. This last factor is of special concern in this paper. Minority groups—such as women in engineering—have less access to role models, with whom to identify personally and socially. Majority groups, such as white men, have the opportunity to interact informally and regularly with various individuals that may serve as mentors to them (Whittaker & Montgomery, 2012). In the case of women in STEM, these interactions often need to be intentional because there are fewer opportunities for them to occur spontaneously.

Mentoring is often used and recommended to attract and retain women in STEM programs (Dennehy & Dasgupta, 2017; Hernandez et al., 2017; Stelter et al., 2021; Stoeger et al., 2017; Ziegler et al., 2019). In fact, given that retention rates in engineering careers are traditionally low, mentoring programs have been recommended regardless of student gender (Lim et al., 2017). In the case of women in engineering, the potential benefits of mentoring may include addressing risks common to students of all social groups as well as those more specific to female students, such as the aforementioned access to role models in engineering.

With this in mind, we designed a mentoring program for female students at the Industrial Engineering School at the Pontificia Universidad Católica de Valparaíso (PUCV) in Chile. The program was planned for May 2021 and it came into being in June of the same year. Our goal was to facilitate the development and strengthening of our first- and second-year students' professional identities. Recent advances in the theory of professional development in STEM suggest that students' professional identity is critical to their motivation toward their career and to improving retention rates. A solid engineering identity means that students have come to see themselves as engineers and have a sense of belonging to the discipline community (Hernandez et al., 2017).

One of our concerns about potential difficulties for running this program was consistent with what is reported in the literature on mentoring minority groups in STEM: the availability of mentors from those underrepresented groups (Whittaker & Montgomery, 2012). Female incoming students to engineering fields will typically find few upper-level female students, few female faculty, and few female graduates.

The motivation for this study originally derived from this concern, and we deemed it important to understand the motivations of mentors, the benefits they perceive from mentoring, and the roles they play for their mentees, in order to successfully strategize and design our mentorship program. However, as the reader will notice, during the course of our study, we enhanced the scope of our analysis from addressing perceived benefits and motivations to now focus on ways of making sense and giving purpose to their involvement in mentorship. With this shift, we now also address mentorship more fully as a practice: a set of social relationships and interactions in a community, which revolve around a good that is produced as a result of those interactions and whose standards of excellence are defined within the community itself (Macintyre, 1984). Of course, someone can be motivated to engage in practice for instrumental reasons; for instance, when one obtains some remuneration for doing so. But it is when the production or achievement of these practice-defining goods becomes a motivator for someone to engage in that practice that it becomes a source of meaning and sense. In the case of mentorship, more particularly, we identified these goods as being characterizable in terms of caring, as understood in the ethics of care framework (Gilligan, 1993; Noddings, 2012; Tronto, 1998; Tronto & Fisher, 1990). As such, it provided an appropriate frame for interpreting the motivations for senior (upper-year) students and graduates (alumni) to become and remain mentors.

Therefore, the purpose of this study is to inquire about how mentors in the Industrial Engineering mentorship program for female students at PUCV make sense of their mentorship practice.

8.2 Mentoring Female Engineering Students

Mentoring broadly indicates a relationship in which a more experienced person—the mentor—provides advice to a less expert person—the mentee (Meschitti & Smith, 2017). Mentoring, defined as a relationship of professional assistance between the mentor and the mentee, is fundamental for the professional development and well-being of the mentees, as well as for the promotion of diversity and equal access in engineering education (Long et al., 2018). Alternative definitions focus on the relationship between mentee and mentor: "a developmental relationship that is intended to support the growth of the learner or protégé forward, involving a constellation of mentors with different assets and skills" (Mondisa et al., 2021).

Mentoring may be implemented across all academic trajectories, stages of development, training opportunities, and career pathways (Meschitti & Smith, 2017; Stelter et al., 2021). Some programs focus on academic support, research training, social integration, or professional identity development. Some programs are formal, with clear expectations, evaluation, or feedback. Others are informal and flexible. Mentoring programs may be voluntary for mentors or mentees, or mandatory. Mentors may be assigned, or they may select their mentee. Mentorship may be carried out in a one-on-one relationship or in group settings, with one or more mentors at a time. They may be face-to-face or virtual. Mentors may be peers (slightly older or more experienced students or colleagues) or senior members or an organization (i.e., professors, managers, industry professionals).

One common aim of many mentors in engineering programs is to attract and retain students and professionals to the discipline, be it high school students considering STEM careers, university students unsure of their career path, graduate students selecting their research area, faculty on tenure track (Meschitti & Smith, 2017) or junior professionals seeking career advancements in an organization (Allen, 2003).

For women in male-dominated fields, research indicates that broader networks of mentors are associated with greater benefits for mentees (Hernandez et al., 2017). The success of STEM mentoring programs for underrepresented groups depends on the engagement of students and the commitment of mentors (Whittaker & Montgomery, 2012). We examine next the mentorship practice focusing on characteristics of the mentor.

8.2.1 Mentor Roles

Among the three main functions of mentoring are career development, psychosocial support, and role modeling (Long et al., 2018). However, each stage of a person's development may require different roles and attributes from the mentor (Allen et al., 1997).

In the mentoring of engineering students, mentors challenge their students to expand their comfort zone (Meschitti & Smith, 2017), they sponsor, counsel, and help them overcome barriers and plan their careers (Crisp & Cruz, 2009). They also help them develop technical skills (Mondisa et al., 2021) and a sense of belonging. One role that industry professionals provide when mentoring is access to their networks, resources, and position to offer their mentees meaningful professional experiences and opportunities. Women mentees often seek support and guidance on issues that go beyond academic topics (Daniels et al., 2019), and mentors should be able to identify and play this role.

8.2.2 Mentor Attributes

Mentors use their cognitive, relational, and emotional abilities to engage their mentees (Johnson, 2003). The mentoring role requires several personal attributes, including the ability to listen, reflect, tell stories, teach/learn, showcase different career paths and the ability to assist their mentees. Roberts (2000) identifies contingent mentoring attributes such as coaching, sponsoring, and role modeling. Mentors often have a prosocial disposition, with characteristics such as "other oriented-empathy", helpfulness, generosity, and kindness (Allen, 2003). Mentors are proactive, rather than reactive, as mentoring involves the active choice to form a caring bond with others (Mayseless, 2015).

8.2.3   *Motivations*

The motivators to become a mentor are mostly of an intrinsic nature. Self-determination theory postulates that people are intrinsically motivated by competence (a feeling of mastery or perception that growth and success are possible), autonomy (the initiative and ownership of one's actions), and relatedness (a sense of belonging and connection) (Ryan & Deci, 2020). People are intrinsically motivated to purposefully look for opportunities, situations, and relationships in which they can provide care, and experience the intrinsic satisfaction of realizing that their care has benefited others (Mayseless, 2015). Providing care, then, can be considered as a source of motivation for mentors.

Beltman and Schaeben (2012) distinguish several rewards to being a peer mentor, such as opportunity for networking, camaraderie and collaboration, sense of satisfaction for helping people, increased confidence and connectedness, the acquisition or improvement on professional and organizational skills, and the development of confidence or empathy. Benefits from networking, friendships, personal growth, and skill development can be also seen as motivators to mentors. In their study of graduate/postdoctoral mentoring undergraduate students in research, (Dolan & Johnson, 2009) identified two main motivators for mentors: one was instrumental and involved improving their research productivity and meeting the implicit or explicit expectations of the research group, and the other was socio-emotional and was related to personal enjoyment and a desire to share expertise.

Allen (2003) identifies similar motivators in workplace mentoring: helping others succeed in the organization (benefiting others), a sense of pride from mentees accomplishments (personal satisfaction), learning, and career success (self-enhancement). Managers with previous experience as mentors tend to volunteer again, but when deciding whether to mentor or not, they evaluate the cost of mentoring (time and personal energy invested) (Malota, 2019).

8.3   Case Study Overview

Engineering programs in Chile range from 10 to 12 semesters long and grant both a Bachelor of Science in Engineering (after completion of the 8th semester) and a professional title. Students choose their engineering specialty early on (often when they apply to the university) with insufficient information regarding potential career paths. One established reason for student dropout and for the gender gap in attraction and access to engineering programs is the lack of professional identification. Although upper-year female students and female faculty contribute to a positive academic experience and are great role models for younger female students, engineering graduates that work in the industry provide better examples of professional career paths and are helpful in developing students' engineering identity. Although the coordination of face-to-face mentorship meetings with engineers working in the field is often difficult, the much higher availability of mentors for online meetings compensates for the loss of intimacy or closeness.

While there is abundant literature on the importance of the connection between mentors and mentees (Dennehy & Dasgupta, 2017; Hernandez et al., 2017), and on the impact of such programs on mentees (Budny et al., 2010; Dennehy & Dasgupta, 2017; Hernandez et al., 2017), research on the benefits for mentors is scarce (Allen, 2003; Beltman & Schaeben, 2012; Rangel et al., 2021; Stelter et al., 2021).

Thus, we wanted to investigate the motivations mentors have when volunteering for an online mentoring program. We study two types of volunteer mentors, peer mentors (senior students) and industry professional mentors (female engineering graduates). The mentees in this program are first- and second-year female students at an engineering school.

This study centers on an online mentoring program (using video call and chat rooms) for female engineering students. It is a 2-phase mixed study addressing mentor motivations and benefits. In the first, quantitative, phase we analyze mentor responses to a 28-item Likert-scale questionnaire that enquired about the importance of motivations for mentoring. We then used the results of this quantitative phase to identify relevant issues that were worth exploring deeper. In a second, qualitative phase, we set up focus groups where some of the program mentors interpreted those relevant issues and connected them to their own personal experiences, exploring what made mentorship a meaningful activity for them to be involved in.

The setting of this study is the Industrial Engineering program at the Pontificia Universidad Católica de Valparaíso in Chile. While the University has 17,000 students, 43% of whom are women (29% women in STEM programs and 57% women in non-STEM programs), the industrial engineering program houses 1000 students, of whom 33% are women, a percentage that has not increased in the last 10 years.

A call was sent via internal networks to senior students and industrial engineering graduates (alumni) to volunteer as mentors during one semester. Simultaneously, we invited first- and second-year female students in industrial engineering to register for the mentorship program. A total of 56 mentees, 22 senior student mentors, and 18 alumni mentors were enrolled. These participants were assigned to 8 groups composed of 4 to 6 mentors and 6 to 8 mentees. In each group, the team of mentors was composed of both graduate mentors and senior student mentors.

Prior to starting the program, we held a mentor training session and provided participants with a guide that compiled tips and good practices for mentoring sessions. Mentors were, however, enticed to adapt this material and design session to their own liking, as the main objective was to engage students in conversation and answer their questions. Senior student mentors served as coordinators, setting up different informal chat groups to facilitate both the task of organizing the session with their team of mentors and the meeting date with their group.

8.4 Looking into Mentor Motivations

The research results of Beltman and Schaeben's study (2012), which looked at peer mentoring of university students across all academic programs for both men and women, served as the basis for the first phase of our study. Beltman and Schaeben (2012) addressed the anticipated benefits of being a mentor, as stated by students when they apply to become mentors. They analytically identified four categories of self-reported benefits by mentors: Altruistic, Cognitive, Social, and Personal Growth. Roughly, altruistic benefits are related to the enjoyment and satisfaction of helping people. Cognitive benefits include learning new information, acquiring new skills, or improving existing skills. Social benefits are related to networking, interacting with other students, or developing friendships with them. Personal growth benefits involve self and personal development and mentors' reflections on their own experiences and growth.

These categories and subcategories were the basis for our mentor questionnaire. We translated them and constructed 28 items (available in Table 8.1), each of them consisting of an assertion whose relevance respondents were asked to rate as factors that contributed to their decision to become mentors. We had 17 responses out of 22 student mentors, and 11 responses out of 18 graduate mentors.

The reliability measurement (Cronbach's Alpha) obtained was 0,909, which suggests that these items may be close to each other in meaning. But this closeness does not mean that all of them point in exactly the same direction. As we wanted to reveal their inner structure, however, not analytically as in Beltman and Schaeben's study, but in the perspective of the program mentors, we carried out an analysis of principal components. Indeed, it did not support the four categories of motivators (benefits) proposed by Beltman and Schaeben (2012) in their study. Instead, as shown in Table 8.1, we found two principal factors (components) which seem to relatively neatly distinguish self-focused motivations (e.g., development of self-confidence, social skills or improvement of their CV) from other-focused motivations (e.g., sharing their experience, inspiring other women, being available for the needs of others, satisfaction for helping others) (Allen et al. 1997). Only one of these items does not appear to fit this characterization (Training to be a mentor seems interesting and relevant).

The reliability within component 1 (self-focused motivations) was very high, with Cronbach's $\alpha = 0,925$, as it was within component 2 (other-focused motivations), with Cronbach's $\alpha = 0,844$. But there seems to be a clear difference between the two components, both analytically and in the results obtained.

Table 8.1 Principal component analysis for mentor motivations

	C1	C2	Uniqueness
Being a mentor allows me to enrich my curriculum	0.790		0.374
Training to be a mentor seems interesting and relevant		0.766	0.356
As a mentor, I develop interpersonal skills	0.833		0.302
Mentoring is an opportunity for personal growth	0.784		0.364
I believe that I can have an impact and make a difference		0.424	0.795
Mentoring allows me to develop communication skills	0.881		0.205
Mentoring allows me to develop leadership skills	0.776		0.398
Mentoring will help me develop empathy	0.402		0.838
Mentoring has a social sense for me			0.906
I am proud to participate in my mentees' achievements		0.774	0.379
Being a mentor allows me to update my knowledge of the University and the School		0.518	0.633
Being a mentor connects me with my School / University	0.594		0.594
I feel satisfaction in helping students integrate		0.719	0.473
I believe that I can share my experience and knowledge		0.866	0.248
By being a mentor, I will be able to coordinate activities with other people	0.711		0.350
Being a mentor contributes to my professional development	0.817		0.333
Mentoring allows me to meet and interact with engineering students			0.770
Mentoring allows me to inspire other women		0.862	0.257
By being a mentor, I develop social skills	0.754		0.430
Being a mentor allows me to participate in a network with other mentors	0.598		0.540
I appreciate receiving feedback from mentees		0.633	0.595
I experience a sense of responsibility for participating as a mentor		0.607	0.569
I like to be available when they need me		0.433	0.754
Mentoring allows me to develop a rewarding relationship with the mentees		0.647	0.513
I would have appreciated a mentoring activity when I was a student	0.683		0.529
I feel pride or a sense of accomplishment in participating as a mentor			0.755
Being a mentor allows me to enhance my coordination and organization skills	0.806		0.347
Mentoring will help me build my confidence	0.675		0.435

When using the calculation of each of the two components as an index, with C1 as a measure of relevance for self-focused motivations and C2 for other-focused motivations, the difference becomes apparent. Beyond the fact that the differences between the means of the two components were *statistically* significant (with p-values <0.001 for t-Student, Wilcoxon and Z tests), they are *substantively* significant. Remarkably, the mean difference (0.593) corresponds to almost 15% of the theoretical range for these variables (1–5). Respondents were more inclined to take other-focused motivations as more relevant for explaining their decision to become mentors than self-focused ones.

This distinction between the two components can be further elaborated. It can be seen in Table 8.1 that the items that contribute to C2 are generally much more intrinsically and directly related to the purposes of mentorship; that is, the relationship between them and mentorship is not contingent but, instead, is central to the goods that are pursued by the practice of mentorship. On the contrary, the items that contribute to C1 are generally much more contingently connected to the practice of mentorship, and the benefits they refer to might rather easily be obtained by getting oneself involved in other, different, practices. For instance, enriching one's *curriculum vitae*, developing social skills, and building up one's confidence can be side results of many other activities completely unrelated to mentorship and can hardly be seen as what this practice pursues. But sharing one's experience and knowledge with others, inspiring other women, and being available for others when they need it is much more clearly connected to what makes mentoring the practice that it is.

Interestingly, although for both groups of mentors the other-focused component is more relevant than the self-focused component, this difference is markedly larger in the case of graduate mentors as compared with senior student mentors. Additionally, in the self-focused component, statistically significant differences were found between both groups of mentors. On the other hand, for the other-focused component, the results for both groups are similar (Table 8.2).

Additionally, a Mann–Whitney U test for comparing the two groups (student mentors and graduate mentors) shows a p-value of 0,007 for the differences between the two components. This indicates that, in fact, graduate mentors mark a much clearer difference between the two components.

It may be perhaps surprising that senior students seem to be similarly motivated by the other-focused component to become mentors when compared with graduates. The

Table 8.2 Descriptive statistics of components by mentor type (student/graduate)

	Component 1		Component 2	
	student	Graduate	Student	Graduate
Valid data points	17	11	17	11
Mean	4.206	3.643	4.556	4.612
Std. Deviation	0.630	0.757	0.476	0.357
Minimum	2.500	2.214	3.545	3.909
Maximum	5.000	4.643	5.000	5.000

fact that they had much more recently been junior students themselves could justify an expectation for them to be more empathetic toward their mentees than graduates who had gone through that experience much longer ago. In regard to the significant difference between the two groups of mentors—concerning their motivations based on expected individual gains/benefits—it can be speculated that this difference could be a result of the broader take on life and work of graduates when compared to senior students, most of whom have not yet had the opportunity to work. Additionally, many of the skills represented in the items that contribute to C1 may refer to competencies that graduates have already developed or that they can develop in their workplace.

Besides the fact that the items contributing to C2 are for the most part directly connected to the goods that define the practice of mentorship, when looked at closely, and as we will explain in more detail in the following section, these other-focused motivations are at the same time instances of *caring*. And that is what we will now turn to.

8.5 Caring

Caring is an emotion, a bond, and a behavior that manifests everywhere around us (Mayseless, 2015). It can be understood as "a species activity that includes everything that we do to maintain, continue, and repair our 'world' so that we can live in it as well as possible. That world includes our bodies, our selves, and our environment, all of which we seek to interweave in a complex, life-sustaining web" (Tronto & Fisher, 1990). As such, these motivations are aligned with a way of seeing the university more as a site of personal relations of solidarity that are aimed at repairing and maintaining *our* world, than as a site of transactions that different actors conduct while each of them seeks their own benefit, as, for example, in a market. All this suggests that mentorship can appropriately be taken as a caring practice.

According to Noddings (2002), care is basic to human life. "Natural care", therefore, is a moral attitude: a longing for the goodness that arises from the experience or memory of being cared for. Noddings explores the notion of ethical care: a state of being in a relationship, characterized by receptivity, affinity, and fascination. Receptive attention is an essential characteristic of an affective encounter. The caregiver is open to what the cared-for is saying and what they might be experiencing, and he/she is able to reflect upon it. However, there is also a motivational drive. In other words, the "motor energy" of the caregiver flows into who is cared-for. The caregiver thus responds to care in a way that is hopefully helpful. For this to be called "caring", it also requires some recognition from those receiving care. Caring implies a connection between caregiver and cared-for, and on a certain degree of reciprocity; that both benefit from the encounter in different ways and both give something to the relationship.

The ethics of care emphasizes that caring relationships between humans are part of what marks us as human beings (Gilligan, 1993; Noddings, 2002; Tronto, 1998; Tronto & Fisher, 1990). We are always interdependent beings. Throughout life, human beings need and receive care and take care of others. There is a motivational aspect to ethics of care-related actions (care as a response to a feeling) and there is a cognitive aspect that arises from the memory of having been cared-for (Torras, 2013). According to Tronto and Fisher (1990), caring can be seen as a process with four intertwined phases: caring-about, caring-for, caregiving, and care receiving (see Table 8.3). Viewing care as a process stems from the fact that each phase feeds the next. However, they should not be seen as standing in a strict sequence, as there is always a toing and froing between them. Moreover, these phases can either be performed by a single person or divided between different individuals or groups. Table 8.3 further develops these phases of care and the moral element or dimension that each involves.

Once we realized that one of the factors found in the principal components analysis pointed more directly in the direction of the goods pursued by the practice of mentorship and that these goods could be characterized as care, we decided to put care right in the center of the framework for our analysis of the qualitative data collected by means of the focus groups. That is what we present in the following section.

Table 8.3 Tronto's four phases of care

Phases of care	Definition (Riley et al., 2017)	Moral dimension (Tronto, 1998)	Care responsibilities (Tronto, 1998)
Caring about	An attentiveness or sensitivity to opportunities to care for others	Attentiveness: perceiving needs in self and others with as little distortion as possible	Listening to articulated needs Recognizing unspoken needs Distinguishing among and deciding which needs to care about
Caring for	The assumption of responsibility to care for another	Responsibility: defining what to do about the needs that were identified	Organizing Marshaling resources or personnel Paying for the care work that will meet the identified needs
Caregiving	The act of meeting another's needs	Competence: knowledge about how to provide good care	Performing the necessary caring tasks Making sure the care is done well
Care receiving	Response to care returns to the caregiver	Responsiveness: assessing reactions of the care-receiver regarding the care received	Listening/Distinguishing/Observing responses from care-receiver Eliciting feedback Identifying new needs for care. This comes full circle to the attentiveness phase (caring-about)

8.6   Understanding How Mentors Care

In phase 2 of our study, we organized 4 focus groups, two for student mentors and two for graduate mentors. In total, 13 student mentors and 6 graduate mentors participated in these focus groups. During sessions, we discussed some of the results obtained in the quantitative phase by prompting mentors with questions such as "What triggered you to say: 'yes, I want to be a mentor'?, What reflections arose when making this decision?", "How is participating in this experience valuable or satisfactory to you?", "What effect did the sharing of your story have on you?", or "What have you learned about yourself during this process?".

As we will see, the caring nature of mentorship became manifest in the mentors' responses, as it showed to be the source of meaning and purpose for their involvement in the program. Its orientation toward the needs of others, in this case, the mentees, not as something done *upon* them as imposed from the outside, but as being there *for* them, is revealed in the following quote from one of the student mentors:

> In the end, impacting doesn't have to do with changing how they [the mentees] think, or their vision of the future, but simply is related with them not feeling that they are alone in this. SM4.[1]

We have pointed out that in the case of mentoring programs for underrepresented groups in engineering, the opportunities for those who belong to these groups to be mentored are reduced by the shortage of mentors from the same social groups. It is especially important to design and manage mentoring programs with attention to aspects that may affect the motivation of the mentors.

Based on the findings of this study, we will examine and discuss the elements that can affect the motivation of mentors under the scope of the four phases of care.

8.6.1   Caring About

The moral element of the caring-about phase is **attentiveness**. Attentiveness is a practice that develops over time, in the process of acquiring experience in becoming a mentor. Initially, when they get involved in the program, their attentiveness is inevitably a projection from their experience with others in a situation that they judge similar to that of the mentees. Perhaps they have been attentive to others in the past and approach their mentees from this working hypothesis or intuition about the kind of caring they might need.

Mentors refer to their own experiences and to what they have observed with peers or as teaching assistants to younger students. They identify that first-year students need counseling during difficult stages of life, such as in the transition between high school and university.

---

[1] From here onwards SM stands for student mentor and GM stands for graduate mentor.

> When I was a teaching assistant at a [college preparatory] program, I noticed that many high school students felt very lost. I felt that becoming a mentor would give me a similar chance to help. The first years at the university are difficult and many students live them in silence. Mentoring is also a way to give them confidence and share my similar experiences. SM1.

> I was a teaching assistant to first-year students and saw the uneasiness of students who didn't know anyone at school. From my own experience, I know that having someone senior to whom to ask questions helps a lot. GM1.

Sometimes that previous experience of feeling the need to be cared about is their own. There is evidence of a motivating force in mentors, resulting from having survived the difficult moments of the first years of university studies. They feel that sharing these experiences contribute to making this process more enjoyable for other people.

> It is always motivating to be invited to take part in these types of activities [referring to the mentoring program] and to be able to contribute as well. We all go through different stages where we experience discomfort, especially during transitions such as entering a university program. I was very motivated by the possibility of giving back and sharing what I learned, making the path easier for newcomers. SM2.

As the caregiver attends, they are likely to experience a motivational shift (Noddings, 2012). In the mentoring program, this means that the mentor redirects her care efforts toward noticing and understanding the needs and goals of the mentee, no longer as a projection of her previous experiences but in present time, and with real rather than imagined persons.

> I had to update my view on gender issues: for me [when in school] the issue was that women were not being considered, (…) smart topics or technical questions were answered by men (…) but for the girls in my group this is not an issue … for them the difficulties have to do with confidence (…) they see their [male] peers a lot more empowered than themselves, and this is a new problem that we have to work with (…) I realized I was thinking of working on the problems I had when I was in school (…) but that has been updated, and it has been great to learn this. GM1.

> I had already been a mentor to engineering graduates in an organizational context. In this program, the group setting was difficult for me, different from my one-on-one mentoring experiences. In the company there is a strategic plan, common areas and tools that connected me with the mentees. Here in the university the challenges of the mentees are more ambiguous, more personal. GM5.

Initially in the program, mentors connected with the needs of mentees by recalling their own experiences and difficulties as female engineering students. Experiences with choosing to study a traditionally male profession, and with different steps in their career path (selecting a major, finding an internship, joining the workforce, advancing in their career). Once the mentoring relationship began, however, mentors updated their understanding of what mentees needed. Relying on their autonomy and on the ability to perceive what is needed was key to detecting new opportunities for care. A mentor demonstrates empathic accuracy when what she infers of a mentee's feelings and thoughts corresponds to what the mentee really feels and thinks (Ickes, 1993). The concept of empathic accuracy raises the question of whether our knowledge of another person's mind proceeds in an egocentric or detached way (Stueber, 2006).

Care has both cognitive and affective dimensions (Noddings, 2012). Dialog is essential to building relationships of care and trust. "Within a well-established relation, we are more likely to achieve empathic accuracy" (Noddings, 2012). In this way, the mentor is prevented from attributing to the mentees the feelings she had or would have under similar conditions. The ethics of care requires the mentor to think about how the other (mentee) feels, not how they would feel under the same circumstances.

This is similar to what Noddings (2012) asserts about the teacher-carer: a mentor-carer listens receptively to mentees. Mentors may recognize that mentees' concerns/interests differ from their preconceptions. In this case, a conflict may arise if the mentor values the original program design as it aligns with her own past experiences and needs. To respond in a caring way, mentors need time to build a caring and trusting relationship with their mentees. This way mentees and mentors can together choose a suitable course of action to address their needs. In any case, the time spent building the relationship will likely be a worthwhile experience for both parties.

8.6.2 Caring For

The decision to join a volunteer mentoring program, with no compensation, monetary or otherwise, demonstrates intrinsic motivation. Caring-for implies being intrinsically motivated to take **responsibility**. After considering their own experiences, mentors perceive helping mentees to be their moral responsibility. As a program that does not provide monetary rewards, this moral boost is an important motivator for mentors.

> I felt it was my duty to pass my experience on to the younger generations. To tell them not to let stress take over, especially during the first years, and to seek help early on. SM3.

> To me, being a mentor is like volunteering for something great. SM4.

> I turned out all right, but I had a really hard time when I was studying engineering. The mentees may be living the same situations that I had to live. It motivates me to share my experience so that they do not have such a difficult time (…). It is my way of showing solidarity. GM2.

Mentors, who as students did not have the opportunity to be mentees express how useful that experience would have been for them. This reflection creates a sense of responsibility and desire to volunteer as mentors. Conversely, those that were mentored as students recognize how important it was for them—how that person made a difference in their life—and they too wish to pay back by becoming mentors themselves.

> My motivation to join had to do with what I experienced in my first year when I was part of a mentoring program. My mentor then was the same faculty that invited me to this program. Her invitation reminded me of how helpful the program had been for me (…) I thought I could reciprocate as a peer mentor. SM5.

Caring in a mentoring relationship goes beyond feeling with the other; it involves acting on behalf of the other. It involves taking responsibility for the welfare of another (Lawrence & Maitlis, 2012).

Mentors' attendance in the program was nearly perfect, unlike the mentees whose attendance was erratic. Their studies, jobs, and personal life would have been perfect excuses for not participating, especially because they didn't personally know the people they were mentoring. What moved them from a caring-about to a caring-for attitude? When mentors talk about why they decided to take on the responsibility of being mentors, it is evident that they are moved by feelings of compassion and "fellow-feeling for others" (Noddings, 2002). Unlike other programs, this is an online program that was designed to allow for flexibility and considerable autonomy from the mentoring groups, which are conditions that might be motivating mentors to directly execute the care, instead of channeling their caring-about interest through over venues (such as charity work).

8.6.3 Caregiving

Mentors evaluate or discover their **competence** to teach and accompany others; their achievements and experiences substantiate this competence.

> I was unsure if my message was going to be adequate: I believe that it is possible and right to reconcile personal life with academic life, that university is not everything. I wondered if this message was in line with the goals of the mentoring program. But knowing that I would mentor with graduates gave me confidence, I wouldn't be alone, and I could learn from them [graduate mentors]. SM7.

For the proper development of a mentoring relationship, the mentor must perceive herself as competent in her role. The limited knowledge or inexperience regarding what happens during a mentoring session, or the preconception that mentoring is like academic tutoring, made some mentors initially second-guess their ability to make any contribution to first- or second-year engineering students. Once they realized how meaningful it was for their mentees to listen to their point of view or experience, they realized that they were competent to fulfill the role of mentor and they acted.

> Our teachers gave us a structure for the mentoring sessions, but we eventually went off the script. We have been talking with the mentees about vocational issues and concerns, how to deal with failing a course, difficulties finding the first job, etc. They are difficult topics that we have discussed based on our experiences and that have been well received by the mentees. Deviating from the mentoring guide script I think has led us to something even better. SM6.

The mentoring literature describes different attributes of a competent mentor. According to Johnson (2003), the competence of a mentor is more than the sum of several skills; indeed, it is a structure that integrates various virtues, abilities, and focal skills that are deployed to serve the mentee. In this study, we are more interested in the mentors' own perception of competence for the work entrusted and voluntarily adopted, and in their efforts to develop their own competence, than in a repertoire of possible attributes of a mentor as a caregiver. Technically, the mentor needs to have knowledge and resources to do a good job (Zembylas et al., 2014). However, as Tronto points out, competence is not just a technical consideration, it is also the moral dimension of caregiving (Tronto, 1998).

"Providing effective mentoring to others requires a considerable time investment on the part of the mentor and can put the perceived competence of the mentor at risk if the protégée performs poorly" (Allen, 2003).

According to the theory of self-determination, the sense of competence is a relevant element for a person's motivation. Within the existing literature on women in engineering, self-perceived competence has been a recurring theme. Female engineering students rate themselves as less competent relative to their male peers, regardless of their grades (Jagacinski, 2013). While the literature refers to technical competence, this study reveals a similar pattern, as some mentors initially doubted their abilities and competence to mentor. We may also interpret this concern about competence as an indication of caring; that is, the moral element of caregiving in mentorship is reflected on the mentors' attention to their own competence and their efforts to improve it, as the result of their wish to ensure to do a good job because they care about the wellbeing of their mentees.

8.6.4 Care Receiving

The role of the person receiving care is simple and crucial (Noddings, 2012). Demonstrating in some way that care has been received does not mean that the mentee should express gratitude, but rather **responsiveness** (e.g., showing up or asking questions).

In some of the mentoring groups, mentees' attendance was not consistent. For mentors, investing their time and energy in preparing activities and then realizing that there is little response from mentees is demoralizing.

> I wish the mentees had participated more (…) the graduate mentors contribute a lot in the sense of work and we in the sense of the university experience. SM7.

> I thought: ah, the girls are going to love it, they are going to be super motivated, they are going to learn a lot, because who wouldn't want to have this experience during their first years? (…) but participation has not been so high. I like to keep things under control, know how they arrive to the meeting and what they leave with. (…) For me this was very important, it was like a pre-conviction that motivated me to be a mentor. SM11.

On the other hand, it is satisfying for mentors to see signs that their sessions have an impact, even if it is modest, on the lives of their mentees.

> It comforts me to see their faces of relief, that someone understands them, when they realize *it is not me who has a black cloud, it is a normal process, everything happens step by step.* GM4.

> I didn't realize in the beginning how satisfactory it was to help, but later I did (…) because the girls showed me. SM10.

Graduate mentors have a double source of reward as they not only see responsiveness from the mentees, but also from the senior students with whom they team to mentor. Senior students, unlike mentees, tend to speak forcefully about how helpful it is for them to talk to graduates.

> It has been very interesting to meet the graduate mentors and to create ties with my peers in the mentoring team (…) I was worried about finding an internship but talking with them has made me feel I can solve it. I have felt their company. SM5.

> We have all learned a great deal from the graduates, how they solve problems and how they deal with difficulties. It has been very interesting to hear from them that not everything is perfect, that entering the world of work is not necessarily easy. SM12.

> In my case, networking was one of the main motivators. I saw it as a network for work, internships, and all that. SM3.

The expectation of future rewards beyond the mentoring program is also a motivator for mentors. The idea is that just as the mentors now cared for the mentees, if ever the former needs help, the latter or the other mentors will be there to support them.

> What is relevant is the relationship between mentors and mentees (…) I don't imagine a close relationship or a friendship forever, but a nurturing relationship, a relationship that contributes, that teaches you, that leaves reflections and a space of trust… a network for the future, for example. For me it was very important. SM13.

> It was very satisfying to create a community (…) our mentoring was not just for first-year students, but the senior students [mentors] also benefited by sharing their problems and receiving our help (…) We identified with one another (…) I feel that if I ever need something this group is going to be there to help and support me. GM6.

To close the cycle of care, a response is needed from those who have been cared-for. In mentoring programs, mentors hope to see an impact on the mentees. Observing signs of relief, seeing that the students are calmer or have more confidence in themselves are aspects that mentors refer to as satisfactory and that allow them to positively evaluate their intervention. The baseline of the response is that mentees attend the sessions. For the mentors of some groups, the low and irregular attendance of mentees was discouraging.

The mentee's response, or lack thereof, involves evaluating the care provided by the mentors. This responsiveness can be interpreted as the degree to which the needs of the mentees have been satisfied in the sessions and activities carried out in this program.

In the program under study, this phase of care is the one that represented the greatest challenge. The online format of mentoring, which is of great advantage for summoning female mentors, may have played against the engagement of some mentees.

8.7 Conclusions and Recommendations

We undertook this study because, given the suitability of mentoring programs to attract and retain women in engineering, and the difficulty of finding female engineering students or graduates (precisely because they are few), it becomes important to understand what factors affect the motivation of women interested in assuming the role of mentor. In studying those motivators, we discovered that what is more important for the mentors in the program is the satisfaction related to pursuing the goods of the practice of mentorship, which is taken to be fundamentally a caring practice.

The framework of the ethics of care has been widely used in feminist literature as a way of understanding how relationships develop between human beings. This approach emerged as a critique of distant and abstract ethical frameworks that omitted basic relational values such as caring or sympathy. In this study, the ethics of care and its phases allowed us to analyze and understand, from an ethical perspective, what drives mentors to enroll and persevere in a mentoring program for younger female students. Conceptualizing mentoring as a practice of care helped us to reveal the moral dimensions that together drive mentors to move from caring-about the gender gap in engineering to assuming responsibility for a specific other and acting upon it.

But we take it that the ethics of care should not only be useful for interpreting the practice of mentorship, but also for designing and adjusting the mentorship program. For instance, the ethics of care's phases or dimensions may allow us to improve the selection process and the training sessions for mentors, and ensure they are relevant and focused on what really matters in the mentor–mentee relationship. It is about devising a program that is tailored to the needs of the mentee, and about working that disposition in the mentor. This framework directs us, predisposes us, and draws attention to certain topics that are generally not seen by program managers, but are important to mentors. For example, we have learned that it is important for mentors to receive a response from mentees, that mentees participate and show engagement with the program. This is, that participants show up after they said they would, answer emails or chat questions, turn on the camera during the online session, or give opinions during a conversation. After identifying this issue, we can review the design of future programs or implement changes to the ongoing program to help intention this participation without having to enforce it. Examples of such modifications are: signing a participation agreement with enrolled mentees before they are assigned to a mentoring group; sharing with mentors and mentees the expectations and behaviors observed in earlier programs; establishing increasing levels of participation/commitment based on observed behavior or declared interests; deciding that mentorship meetings be held in periods of low academic demands; or defining other means to provide timely feedback to mentors, such as offline or anonymous questionnaires.

Importantly, by understanding mentorship as a caring practice and designing mentorship programs in accordance with this view, we can contribute to establishing the university as a human place and to resisting the advances of a neoliberal culture that have been encroaching on it for the past decades. However, even caring may be marketized, as some authors have argued that has happened in the twenty-first century (Green & Lawson, 2011). The practice of care becomes a tradable asset, where caregivers care under a purely contractual relationship.

When caring is viewed as a commodity, the act of caring comes to be perceived as a matter of choice for the caregiver, who may accept (or not) responsibility for engaging in a dependent relationship with others. In a mentoring program with no external rewards for assuming this responsibility, the mentor voluntarily donates her time and energy to a cause she considers valuable regardless of who the mentees are, and thus all are included. A mentoring program such as the one described in this study distances "care" from a market logic and restores its "inherent to human nature" character.

Lastly, we would like to say that the relational nature of caring in itself suggests that this endeavor is a collective one, involving mentors and mentees as well as faculty. In Tronto's words,

> Caring should take place in an environment in which all of those engaged in caring –caregivers and care receivers as well as other responsible parties– can contribute to the ongoing discussion of caring needs and how to meet them. No single actors in a care process can assert their own authoritative knowledge in the process. Within the activity of caring itself, actors must continue to be attentive, responsible, competent, and responsive to the others in the caring process. (Tronto, 1998).

References

Allen, T. D. (2003). Mentoring others : A dispositional and motivational Mentoring others : A dispositional and motivational approach q. *Journal of Vocational Behavior, 62*, 134–154. https://doi.org/10.1016/S0001-8791(02)00046-5.

Allen, T. D., Russell, J. E., & Maetzke, S. B. (1997a). Formal Peer Mentoring. Factors related to protégés' satisfaction. *Gr Organ Manag, 44*, 488–507. https://doi.org/10.1177/105960119722 4005.

Allen, T. D., Poteet, M. L., & Burroughs, S. M. (1997b). *The Mentor's Perspective : A Qualitative Inquiry and Future Research Agenda, 89*, 70–89.

Beltman, S., & Schaeben, M. (2012). *Institution-Wide Peer Mentoring: Benefits for Mentors, 3*, 33–44. https://doi.org/10.5204/intjfyhe.v3i2.124.

Bonaldi, F., & Silva, F. (2014). Gendered habitus in engineering: Experiences of brazilian students. *International Journal Gender, Science Technology, 6*, 21.

Budny, D., Paul, C. A., & Newborg, B. B. (2010). Impact of peer mentoring on freshmen engineering students. *Journal STEM Education Innovation Research, 11*, 9–24.

Cadaret, M. C., Hartung, P. J., Subich, L. M., & Weigold, I. K. (2017). Stereotype threat as a barrier to women entering engineering careers. *Journal Vocat Behavior.* https://doi.org/10.1016/j.jvb.2016.12.002.

Cech, E., Rubineau, B., Silbey, S., & Seron, C. (2011). Professional role confidence and gendered persistence in engineering. *American Sociological Review, 76*, 641–666. https://doi.org/10.1177/0003122411420815.

Crisp, G., & Cruz, I. (2009). Mentoring college students: A critical review of the literature between 1990 and 2007. *Research in Higher Education, 50*, 525–545. https://doi.org/10.1007/sl.

Daniels, H. A., Grineski, S. E., Collins, T. W., & Frederick, A. H. (2019). Navigating social relationships with mentors and peers: Comfort and belonging among men and women in STEM summer research programs. CBE-Life Science Education, 1–13 . https://doi.org/10.1187/cbe.18-08-0150.

Dennehy, T. C., & Dasgupta, N. (2017). Female peer mentors early in college increase women ' s positive academic experiences and retention in engineering, 114. https://doi.org/10.1073/pnas. 1613117114.

Dolan, E., & Johnson, D. (2009). Toward a holistic view of undergraduate research experiences: An exploratory study of impact on graduate/postdoctoral mentors. *Journal of Science Education and Technology, 18*, 487–500. https://doi.org/10.1007/s10956-009-9165-3.

Falco, L. D., & Summers, J. J. (2019). Improving career decision self-efficacy and STEM self-efficacy in high school girls: evaluation of an intervention. *Journal of Career Development, 46*, 62–76. https://doi.org/10.1177/0894845317721651.

Gilligan, C. (1993). *In a different voice: Psychological theory and women's development.* Harvard University Press.

Green, M., & Lawson, V. (2011). Recentring care : Interrogating the commodification of care. *Social and Cultural Geography, 12*, 639–654. https://doi.org/10.1080/14649365.2011.601262.

Hernandez, P. R., Bloodhart, B., Barnes, R. T., Adams, A. S., Clinton, S. M., Pollack, I., Godfrey, E., Burt, M., & Fischer, E. V. (2017). Promoting professional identity, motivation, and persistence: Benefits of an informal mentoring program for female undergraduate students. *PLoS ONE, 12*, 1–16. https://doi.org/10.1371/journal.pone.0187531.

Ickes, W. (1993). Empathic accuracy. *Journal of Personality, 61*, 587–610.

Jagacinski, C. M. (2013). Women Engineering students: Competence perceptions and achievement goals in the freshman engineering course. *Sex Roles, 69*, 644–657. https://doi.org/10.1007/s11 199-013-0325-9.

Johnson, W. B. (2003). A Framework for conceptualizing competence to mentor. *Ethics and Behavior, 13*, 127–151.

Koul, R. (2018). Work and family identities and engineering identity. *Journal of Engineering Education, 107*, 219–237. https://doi.org/10.1002/jee.20200.

Lawrence, T. B., & Maitlis, S. (2012). Care and possibility: Enacting an ethic of care through narrative practice. *Academy of Management Review, 37*, 641–663.

Lim, J. H., MacLeod, B. P., Tkacik, P. T., & Dika, S. L. (2017). Peer mentoring in engineering: (un)shared experience of undergraduate peer mentors and mentees. *Mentor Tutoring Partnersh Learn, 25*, 395–416. https://doi.org/10.1080/13611267.2017.1403628.

Long, Z., Buzzanell, P. M., Kokini, K., Wilson, R. F., Batra, J. C., & Anderson, L. B. (2018). Mentoring women and minority faculty in engineering: A multidimensional mentoring network approach. *Journal Women Minor Science Engineering, 24*, 121–145. https://doi.org/10.1615/JWo menMinorScienEng.2017019277.

Macintyre, A. (1984). *Tras la virtud.* Editorial Crítica. Grijalbo, Barcelona.

Malota, W. (2019). Why managers want to be mentors? the role of intrinsic and extrinsic motivation and the anticipated costs of mentoring for the propensity to mentor by managers in formal mentoring ... Why managers want to be mentors ? The role of intrinsic and extrinsic. *Journal Management Business Administration Central European, 27*, 64–82. https://doi.org/10. 7206/jmba.ce.2450-7814.253.

Marra, R. M., Rodgers, K. A., Shen, D., & Bogue, B. (2009). Women engineering students and self-efficacy: A multi-year, multi-institution study of women engineering student self-efficacy. *Journal of Engineering Education, 98*, 27–38.

Mayseless, O. (2015). *The caring motivation: An integrated theory.* Oxford University Press.

Meschitti, V., & Smith, H. L. (2017). Does mentoring make a difference for women academics? Evidence from the literature and a guide for future research. *Journal of Research in Gender Studies, 7*, 166–199. https://doi.org/10.22381/JRGS7120176.

MCTCI (2020). Radiografía de género en ciencia, tecnología conocimiento e innovación. Ministerio de Ciencia Tecnología Conocimiento Innovación.

Mondisa, J. L., Packard, B. W. L., & Montgomery, B. L. (2021). Understanding what STEM mentoring ecosystems need to thrive: A STEM-ME framework. *Mentor Tutoring Partnersh Learn, 29*, 110–135. https://doi.org/10.1080/13611267.2021.1899588.

Noddings, N. (2002). *Starting at Home: Caring and social policy*. University of California Press.

Noddings, N. (2012). The caring relation in teaching. *Oxford Review Education, 38*, 771–781. https://doi.org/10.1080/03054985.2012.745047.

Powell, A., Dainty, A., & Bagilhole, B. (2012). Gender stereotypes among women engineering and technology students in the UK: Lessons from career choice narratives. *European Journal of Engineering Education, 37*, 541–556. https://doi.org/10.1080/03043797.2012.724052.

Rangel, V. S., Jones, S., Doan, V., Henderson, J., Greer, R., & Manuel, M. (2021). The Motivations of STEM mentors. *Mentor Tutoring Partnersh Learn, 29*, 353–388. https://doi.org/10.1080/136 11267.2021.1954461.

Riley, D., Pawley, A. L., Tucker, J., & Catalano, G. D. (2017). Feminisms in engineering education: transformative possibilities, *The Johns Hopkins University Press, 21*, 21–40. http://www.jstor.org/stable/20628172.

Roberts, A. (2000). Mentoring revisited: A phenomenological reading of the literature. *Mentor Tutoring Partnersh Learn, 8*, 145–170. https://doi.org/10.1080/713685524.

Ryan, R. M., & Deci, E. L. (2020). Intrinsic and extrinsic motivation from a self-determination theory perspective: Definitions , theory , practices , and future directions. *Contemporary Education Psychology, 61*,101860 .https://doi.org/10.1016/j.cedpsych.2020.101860.

Salikutluk, Z., & Heyne, S. (2017). Do gender roles and norms affect performance in Maths? the impact of adolescents' and their peers' gender conceptions on maths grades. European Sociological Review.https://doi.org/10.1093/esr/jcx049.

Smith, E. (2017). Shortage or surplus? A long-term perspective on the supply of scientists and engineers in the USA and the UK. *The Review of Education, 5*, 171–199. https://doi.org/10.1002/rev3.3091.

Steenkamp, H., Nel, A. L., & Carroll, J. (2017). Retention of engineering students. *IEEE Global Engineering Education Conference EDUCON*, pp. 693–698 .https://doi.org/10.1109/EDUCON.2017.7942922.

Stelter, R. L., Kupersmidt, J. B., & Stump, K. N. (2021). Establishing effective STEM mentoring relationships through mentor training. *Annals of the New York Academy of Sciences, 1483*, 224–243. https://doi.org/10.1111/nyas.14470.

Stoeger, H., Hopp, M., & Ziegler, A. (2017). Online mentoring as an extracurricular measure to encourage talented girls in STEM (science, technology, engineering, and mathematics): an empirical study of one-on-one versus group mentoring. *The Gifted Child Quarterly, 61*, 239–249. https://doi.org/10.1177/0016986217702215.

Stout, J. G., Dasgupta, N., Hunsinger, M., & McManus, M. A. (2011). STEMing the tide: using Ingroup experts to inoculate women's self-concept in science, technology, engineering, and mathematics (STEM). *Journal of Personality and Social Psychology, 100*, 255–270. https://doi.org/10.1037/a0021385.

Stueber, K. (2006). 2 Introduction. In: *Rediscovering empathy: Agency, folk psychology, and the human sciences*. MIT Press, pp. 2–28.

Torras, E. (2013). El aprendizaje colaborativo en línea y la ética del cuidado. *Online Collab Learn Ethics Care, 24*, 149–171.

Tronto, J. (1998). An ethic of care. *J Am Soc Aging, 22*, 15–20. https://doi.org/10.1016/S0002-9343(99)80127-2.

Tronto, J., & Fisher, B. (1990). Toward a feminist theory of caring. In M. K. Nelson (Ed.), *Abel EK* (pp. 35–62). Work and identity in women's lives. SUNY Press.

UNESCO. (2017). Cracking the code: Girls' and women's education in science, technology, engineering and mathematics (STEM).

Whittaker, J. A., & Montgomery, B. L. (2012). Cultivating diversity and competency in STEM: Challenges and remedies for removing virtual barriers to constructing diverse higher education communities of success. *Journal Undergrad Neuroscience Education*, 11.

Zembylas, M., Bozalek, V., & Shefer, T. (2014). Tronto's notion of privileged irresponsibility and the reconceptualisation of care: Implications for critical pedagogies of emotion in higher education. *Gender and Education, 26*, 200–214. https://doi.org/10.1080/09540253.2014.901718.

Ziegler, H. S. A., Debatin, T., & Heilemann, M. (2019). Online mentoring for talented girls in stem: The role of relationship quality and changes in learning environments in explaining mentoring success. *New Directions for Child and Adolescent Development, 2019*, 75–99. https://doi.org/10.1002/cad.20320.

Chapter 9
Gender Perspective in STEM Disciplines in Spain Universities

Encina Calvo-Iglesias, Irene Epifanio, Sonia Estrade, and Elisabet Mas de les Valls

Abstract In this paper we present different initiatives carried out by Spanish universities for the incorporation of the gender perspective in STEM disciplines. One of these initiatives is the collection of guides of the Vives University Network for university teaching. These guides cover the sections of objectives, contents, evaluation, learning environment, organizational modalities, teaching methods, and didactic resources with the aim of making women scientists visible in the discipline and eliminating the androcentric vision that predominates in science and engineering. In particular, we analyze the fields of engineering, mathematics, and physics. With the aim of being more than just a review of different initiatives, the paper unifies the fundamentals on which these initiatives are based. Thus, the general principles are well defined, and those aspects more related to each university and discipline particular cultures are identified. The comparison between initiatives will allow us to identify both successful strategies and resistances. Sometimes, the confluence of different events allows an action to become relevant or not. As a result, the paper can be used to effectively define the implementation strategy of the incorporation of gender perspective in STEM teaching at university level.

Keywords STEM · Gender equality · Higher education

E. Calvo-Iglesias (✉)
Universidade de Santiago de Compostela, Santiago de Compostela, Spain
e-mail: encina.calvo@usc.es

I. Epifanio
Universitat Jaume I, Castelló, Spain
e-mail: epifanio@uji.es

S. Estrade
Universitat de Barcelona, Barcelona, Spain
e-mail: sestrade@ub.edu

E. Mas de les Valls
Universitat Politècnica de Catalunya, Barcelona, Spain
e-mail: elisabet.masdelesvalls@upc.edu

© The Author(s) 2022
F. J. García-Peñalvo et al. (eds.), *Women in STEM in Higher Education*, Lecture Notes in Educational Technology, https://doi.org/10.1007/978-981-19-1552-9_9

9.1 Introduction

Spanish women are majority among university students, but they continue to be a minority in STEM careers (Science, Technology, Engineering, and Mathematics). According to the report *Científicas en Cifras 2021* (Unidad de Mujeres y Ciencia, 2021) the percentage of female students enrolled in undergraduate studies in the area of Health Sciences is 70.8% while in the area of Architecture and Engineering is 25.1%. The intrinsic cause of these gendered preferences or orientations when choosing the studies respond to multiple factors as stated in the work carried out by the GenTIC research group at Universitat Oberta de Catalunya (Sáinz, 2011; Sáinz et al., 2020; González-Pérez et al., 2020). These factors include family and teachers' influence, stereotypes, the perception on maths and ICT abilities, and outstanding personalities in the field, among others. The understanding of such factors can significantly contribute to design efficient approaches to increase girls' engagement to engineering studies. Even more, it elucidates how gender perspective can be introduced in the engineering curricula.

This gender gap, which is also produced in other Western countries, has negative consequences for society and women and reduces their job opportunities in the labor market especially in some sectors such as technology, where more net employment will be created in the coming years (World Economic Forum, 2016). Women's underrepresentation in STEM careers has serious consequences for how research is conducted and applied (García-Dauder & Pérez-Sedeño, 2017; Schiebinger, 2021) . From the educational field, we can contribute to solve this problem by introducing the gender perspective in teaching so that "students learn to problematise dominant socialisation patterns and develop skills that will enable them to avoid gender blindness in their future careers" (Catalan University Quality Assurance Agency, 2018). Gender equality goals for higher education are found in all national legislations. It should be remembered that we have a normative framework for the introduction of the gender perspective in higher education in force at European and state level (Verge, 2021). However, reality shows us that the level of inclusion of the gender perspective is low (Verge & Cabruja, 2017) because as Verge exposes "absence of instructions and lack of supervision of the implementation of the gender mandates by evaluation agencies have allowed universities to ignore and even actively resist the call for curricular reform" (Verge, 2021). Fortunately, the policy innovations recently introduced by the Catalan University Quality Assurance Agency (AQU Catalunya) can be a model to Spanish quality assurance agency (ANECA), thus contributing to integrate the gender perspective in the functioning of higher education institutions and in the design and implementation of their programs (Verge, 2021).

This article examines the similarities and differences in the culture of physics, engineering, and mathematics, showing initiatives that can help bridge the gender gap in these disciplines. Among them, the guides for the introduction of the gender perspective in Physics, Mathematics, and Industrial Engineering that are part of the collection *Guies per a una docència universitària amb perspectiva de gènere* published by the Vives University Network. A pioneering resource in the European

Higher Education Area recognized by the European Institute for Gender Equality (EIGE) is an example of good practice in its Toolkit for Gender Equality in Academia and Research (GEAR).

9.1.1   Common Aspects of STEM Disciplines

A significant proportion of faculty and students believes that inequalities are overcome, and many young women say they have never experienced, or have seen, any discrimination. As a logical result of their identification with the elitism and rationality of the fields, they understand meritocracy as neutral. Van den Brink and Stobbe (2014) describes understanding meritocracy as neutral in terms of the following fallacy: the informal practices that help men in their careers are so normalized in the professional culture as to be invisible, while the formal policies established to help women overcome these disadvantages are conspicuous; thus, what is done openly to encourage women to advance professionally is understood as being unfair by most men but also by an important number of women, especially those women who have built an identity as part of the elite (and whom the system harms as well). As Sara Ahmed tells us (Ahmed, 2017), doing equality work means crashing into walls. Walls that were previously concealed and are painfully revealed through the crashing act itself.

Women in academia in Spain have different workload from their male peers, according to the sample analyzed in the article by Cabero and Epifanio (2021). There is a gender gap about the time spent on quality of teaching and caring tasks. Nevertheless, teaching quality is not nearly taking into account in the national system of evaluation for promoting. The big difference is on the number of hours devoted to care of people and domestic work. During the range from 35 to 40 years, the most frequent period of parenting, the difference between women and men is on average 30 h per week for care.

All these things should be taken into account in equality policies. A recent paper critically reviews the implementation of gender perspective in the regulation of scientific and technological research in Spain (Barrios et al., 2021), based on the document "Stop Discriminación" (https://mym.rsme.es/images/docs/mym/stop.pdf).

Concerning the introduction of gender perspective in curricula, a deep analysis of the work carried out at GenTIC research group at Universitat Oberta de Catalunya (Sáinz, 2011; Sáinz et al., 2020; González-Pérez et al., 2020) yields to the identification of three key aspects and the corresponding needs. First, it is observed that female STEM students show a lower self-confidence than their counterpart what might partially be caused by the perception of a lesser ability in maths and ICT during their high school studies. Such perceptions persist during their university studies. Thus, the empowerment of female students is a need. Second, there are still gendered stereotypes in our society and, specifically, concerning STEM occupations. Thus, stereotypes need to be revised and awareness must be raised regarding the consequences of gender biased stereotypes. Third, the lack of female outstanding personalities in STEM field prevents female students to be engaged. Thus, female referents within the STEM field need to be included. The gender perspective in engineering disciplines implies meeting these needs.

In the following lines, the main issues and most interesting strategies identified all along a pilot project at the UPC (Peña et al., 2021) are summarized, following the four pillars classification according to Catalan University Quality Assurance Agency (2018) (i.e., methodology, learning environment, contents, and assessment).

Introducing gender perspective in the methodology is intrinsically related to co-education. This implies active methodologies such as project base learning (PBL), cooperative activities, and debates among others. However, including such methodologies does not guarantee that gender perspective is considered, but it depends on its implementation. Nevertheless, active methodologies can be an efficient framework where gender activities can be easily included, the idea is to reduce competition and emphasize collaboration, and all this is combined with increasing confidence in one's own abilities and reducing the threat of stereotype, as well as improving the sense of belonging. In the frame of the European Higher Education Area, active methodologies have been gradually introduced in the Spanish STEM curricula. Despite this, in the majority of physics, engineering or maths subjects, gender perspective has not been integrated nor made explicit. Moreover, the usage of non-sexist nor andro-centric language is still missing. An interesting approach to co-education and its characteristics can be found in Gónzalez-González et al. (2019).

All of these active methodologies involve peers' interaction what corresponds to the second pillar of the curricula, the learning environment management. Here, deep work is being performed by gender-sensitive teachers when dealing with teamwork. The explicit analysis in the classroom related to gender distribution among groups and the roles and their gender biases, the rotative nature of such roles and special focus on leadership are feasible activities for STEM subjects.

The learning environment management pillar also includes students' interaction with the teaching staff both inside and outside the classroom. Also, it has been observed that female students feel significantly less comfortable than male students participating in the classroom, whereas out-of-the classroom the differences persist but are less significant (Alsina et al., 2019). In order to guarantee that female students equally participate to the learning process, actions promoting their participation by empowering them and increasing self-confidence are slowly being implemented.

The third pillar, contents, is the one that represents the main obstacle for most of the teachers that start introducing the gender perspective in their STEM subjects. Identifying specific subject contents related to gender is not straightforward in most of the cases. However, each STEM subject provides a good opportunity to visualize female engineers and scientists. Indeed, in the analysis of the questionnaire answered by students at UPC during 2018–2019, it is observed that there is a lack of female referents, especially those internationally recognized in engineering fields (Alsina et al., 2019). This scenario is of major concern since the female referents can strongly condition the professional expectations of our female students (Botella et al., 2019). There are some initiatives among STEM universities in Spain designed to increase the number of female referents, and an interesting example is to propose to students the creation of a female biographical profile on Wikipedia (Calvo-Iglesias, 2020).

The remaining fourth pillar of curricula is the assessment. Gender biased peer-assessment and teachers' assessment have been widely reported (Rasooli et al., 2018; Hofer, 2015). Also, several studies can be found related the lower qualifications obtained by female students in multiple choice tests (Birenbaum & Feldman, 1998; Pekkarinen, 2015; Riener & Wagner, 2017), highlighting the fact that not only the methodology but also the instruments are to be deeply revised. However, major concern is focused on how the assessment of the new gender-related actions has to be defined and described in the teacher guide. Indeed, teacher guides in STEM studies are starting to include some of the above mentioned aspects, especially in those universities where a gender dedicated transversal competence exists and hence, not only the competence but also the learning results have to be explicitly introduced.

9.1.2 Main Differences Among STEM Disciplines

The objective of study of physics is what happens in the material world, what we can observe experimentally, formalize in abstract ways, and from these abstractions, make predictions about future experimental observations. Following the ideas of Barad (2007), we understand that in experiments, in measurements, a reality materializes that did not exist before (because it was not known). Although the answer given by nature is impossible to deduce a priori and cannot be built from any discourse, the question asked, how the experiment is designed, what phenomenon is considered important to study, this contains the whole discursive burden of the society that produces that knowledge. Furthermore, this response is also articulated in a discursive way, with metaphors and interpretations that respond to a contingent cultural intelligibility.

Physics has traditionally been done by men, with genealogies of men and stereotypically masculine values. In this sense, Keller (1985) tells us that insofar as physics is defined by those who have practiced before it and practice it now, anyone who wants to enter this community must adapt to existing codes; anyone who does physics without being a man, therefore, has, at the same time, more difficulties but also more potential to subvert them.

The formalism and abstraction of physics can make us lose sight of the fact that when we do physics, we are doing gender. In fact, scientific production, because of its system of rewards through funding, publications, citing practices, etc., is embedded by a patriarchal logic. In addition, science is communicated, also in class, through androcentric coded discursive practices.

Another very prevalent belief, according to Sainz et al. (2017), is that people who engage in physics are "geeks" and have few social skills, but instead are very intelligent. This profile does not align with the gender roles traditionally assigned to women, as they are expected to like (and be competent in) social interaction. In addition, gender stereotypes lead to significant biases among teaching staff (Carlone,

2004). Active teaching methodologies do not appear to have a positive gender impact either (Brewe et al., 2010). Research indicates that even those women who choose to study physics find it difficult to reconcile their being women with a legitimate identity as scientists (Danielsson, 2012; Gonsalves, 2014). It is no wonder women continue to be underrepresented in physics at all levels (Sistema Integrado de Información Universitaria (SIIU), 2019; Figures, 2019), despite decades of interventions of various kinds (Archer et al., 2020).

Physics is taught as an authoritative, unquestionable set of knowledge that is transmitted in such a way that it can maintain a high level of difficulty and status (Hughes, 2001). Students arriving at the faculty must negotiate these dimensions of elitism and difficulty by confronting them first of all with their own coordinates of gender, class, parents educational level, ethnicity, and so on, and how they relate to stereotyped ideas of what kind of person is brilliant in this field. Then, the pace and level of abstraction of the studies themselves, despite being a wonderful way to access a world of ideas and information that can be extraordinarily enjoyable (when the student finish their studies, they usually miss learning so many interesting things in such a short time), is also a constant threat (and in many cases an erosion) to theirself-image as people capable of doing physics. In this sense, the feeling of belonging is reinforced by codes of behavior and expression, very close to the "geek" culture, which are also very much gendered.

In particular, the jokes and memes that form the backbone of this membership, and which are shared by both teachers and students, have the apparent goal of making people laugh, but actually serve to establish or strengthen ties within the community, while excluding those who do not belong, that do not understand the joke or do not find it funny (Johansson & Berge, 2020). And here, in this grouping and excluding, the hidden gender curriculum is incorporated in a particularly insidious way, as it is very difficult to look critically at the sexism implicit in a social practice, these jokes, where adherence or not to enthusiasm and laughing is a kind of shibboleth, a password, which indicates whether or not a given individual is part of the group.

Contrary to Physics stereotypes, the Engineering field has been generally associated with a rich labor market, social recognition, and success (Ariño et al., 2019; Sáinz et al., 2020). Paradoxically, despite such positions require a high degree of communication skills and teamwork, the stereotypical profile of an engineer is someone with a high level of individualism and with a preference on the object oriented work. As an example, the key attractions for engineering students have traditionally been cars, motorbikes and space rockets, among others. Many projects and competitions exist related to these technologies among engineering universities. This scenario prevents girls to be engaged in engineering studies. Moreover, it significantly reduces the sense of belonging to the already engaged girls. Indeed, the social utility value was the main motivation of already engaged STEM girls (Sáinz et al., 2020), being such utility concept mostly related to health and environmental problems solving. It is now the time to change the focus of the engineering interest to problems related with the 17 Sustainable Development Goals, not only in order to increase girls' engagement, but to change the goals and stereotypes of all, male and female, engineering students and to, gradually, contribute to a change in our society.

It is worth to mention that the revision of the stereotypes related to the engineer (i.e., aggressive leadership, high self-concept, business oriented, ...) and the reorientation of the engineering objectives toward social utility values also allow male students to find to define their gender identity, out of the traditional roles and limitations.

In math, we can find now both stereotypes. One the one hand, we find the "geek archetype" of solitary persons who are proud of being very abstract and not seeing any application to what they are doing nowadays. On the other hand, mathematicians are now in high demand in business due to the big data and artificial intelligence applications, which fits the "technological archetype". None of them fits the stereotypical female role.

9.2 Key Actions

The equality committee of the Physics Faculty of Universitat de Barcelona has carried out several initiatives to generate an understanding of the very fact that when we do physics, we do gender. If we are not aware of this, if we carry on with business as usual, we are doing gender in a patriarchal way. We do not advocate to include gender in physics (it is there), we do not attempt to teach feminism (feminism cannot be taught). We want to give the tools, to our students and colleagues, to reach an "aha" moment and start seeing the patterns of exclusion in the codes we have to learn and internalize to become physicists.

We have not been alone in this, a summer school on physics and gender was designed and implemented in the context of the European project Diversity in the culture of physics (Erlemann et al., 2019), and now the project is over will be continued as a Catalan summer school co-organized with Universitat Autònoma de Barcelona, Universitat Politècnica de Catalunya, AMIT-Cat, and SCF. This summer school is offered to women finishing their degrees so that students can network with each other, learn about which research fields are open for them to carry out a Ph.D., how a career in physics looks like, gender as relevant to physics, and available resources for equality advancement.

Besides the summer school, the equality committee of the Physics Faculty of Universitat de Barcelona also carries out activities for the students to understand the matrix of privileges or lack thereof in which they navigate or to critically review the metaphors used by famous physicists. In addition, actions are carried out to try to reverse the epistemic injustice and to recover genealogies of female physicists of the past for the canon. Seminars are offered that help incorporate gender perspective into research, entrepreneurship, and dissemination, especially for doctoral students, and we also offer seminars to teaching staff, where the importance of the discourse is emphasized, not just in terms of inclusive language, but also with regard to the examples or situations described in the problems, and the jokes which are made in class.

In order to spread how to include the gender perspective to math not only to university teachers but also to teachers of all educational levels, a virtual asynchronous math coeducation course (http://www.coeducamates.uji.es/) was carried out in November 2020 at Jaume I University. Being free, recognized by educational authorities as official training, virtual and asynchronous made that participation was very high: more than six hundred people from all over Spain took part. Being asynchronous was key because it favors conciliation. The course was formed by short videos of more than 8 h of duration in total, and written material. There were (written) forums for each area and debates where very profitable, with thousands of comments. In the aim to assess that participant met the course learning requirements, a multiple-choice test was arranged at the end. The resources of the course are available by free for everybody, even not being participant. More details about the usefulness and assessment of this course by participants can be found in Epifanio et al. (2021).

Two complementary strategies can be adopted to include a gender perspective in curricula. The first consists of including specific subjects, for example, the subject *Gender Relations, Science, Technology and Society* offered by the University of Valencia in the degrees in Science and Engineering and Architecture, and in the degree in Pharmacy or the subject *Gender and Science* in the new degree in Science, Technology and Humanities taught jointly by the Carlos III University, Autonomous University of Madrid and the Autonomous University of Barcelona. The second strategy is the introduction of gender perspective in curricula through a cross-sectional approach. In this approach, gender dimension can be included in all subjects through the four pillars of curricula (methodology, learning environment, contents and assessment) described above. It implies a cultural change not only in the institution policies but also in each one of the members of the teaching staff. To this aim, training courses on gender in teaching are being conducted all along the Spanish universities including the strongly masculinized engineering ones. However, since several resistances might be found during this cultural change, as identified in a recent article (Linkova & Mergaert, 2021), more proactive actions must be defined. An interesting example is the pioneer pilot project held at the Universitat Politècnica de Catalunya (UPC) during the course 2018–2019 (Peña et al., 2021). It was a volunteer project where 35 teachers (40% men, belonging to 8 Bachelor and Master Degrees) reviewed the state of the art on gender in teaching, discussed strategies, experiences, and points of view and defined new activities to be implemented during the project. The final outcome of the project was the UPC guide, designed as a checklist, with the aim of becoming an easy and practical tool for the whole UPC teaching community. In the definition of this pilot project there was the clear intention to provide a new learning experience, with less gender biases, to the UPC students all along their studies. To this aim, each of the 8 participating degrees was represented by a team of at least 3 teachers, with subjects at different courses to provide this feeling of coherence all along the studies.

Similar innovative projects are being currently performed in other Spanish universities. For instance, a pilot plan designed for the Electronic Engineering Degree of Universitat de Barcelona from 2018 and funded through a PEVG-2019 project (Estradé, 2021) contemplates gender mainstreaming in all courses, with a strong focus on teaching staff training. Also, specific gender content is introduced in some subjects (Laboratory Fundamentals, Economy, and Projects), and an optative course on Science, Technology and Gender is offered as well.

9.3 Guides for Mainstreaming Gender in University Teaching

Guides for mainstreaming gender in university teaching are a set of guides covering various disciplines that provide recommendations for regendering course goals and contents, references, and teaching and assessment methods. The goal "is to erode the professoriate's resistance to gendering courses, compensating for its lack of gender training, and combating the belief that gender is not applicable to certain fields with practical examples on how to do so" (Verge, 2021).

9.3.1 Physics

"The subject of all physics is affected by the background of the researcher, teacher, and student, and it follows that a gender perspective is needed" (Brage, 2019). And to do this, it is key to publicize the good practices that have been carried out in this subject (or others in the same field) in both Spanish and foreign universities. This is what is intended in the physics guide (Calvo-Iglesias, 2021) that covers all aspects (competences, methodology) showing how the gender perspective can be introduced in physics.

This guide, which is available in four languages, has been presented at various training courses for university teachers organized by the Universities of Alicante, Valencia, Girona, and Extremadura. It has also been part of the webinars organized by the Specialised Group of Women in Physics on the occasion of the 11th February (http://www.gemf-rsef.es/11-de-febrero-2021/) and has also been presented at other conferences.

9.3.2 Mathematics

In the mathematical guide (Epifanio, 2020), a lot of emphasis is placed on the humanization of problems, not only for social justice but also for breaking with the stereotype that math is not useful for real life.

In basic statistics we can use real data to introduce gender biases, such as the Jennifer-John effect (Moss-Racusin et al., 2012) or the importance of sex and gender analysis in science and engineering (Tannenbaum et al., 2019), but social justice problems can be introduced in other math subjects (Buell & Shulman, 2019).

Math guide also highlights the importance of inclusiveness of minoritised groups. It shows that LGBTQ content can be included in math problems (StoneWall, 2017) and it provides resources for the visibility of LGBTQ people, people with functional diversity or other people who do not fit the archetype of mathematician.

9.3.3 Engineering

The concepts and strategies included in the Industrial (or Mechanical) Engineering guide (Mas de les Valls & Peña, 2020) can be easily extended to other engineering studies. In the guide, a detailed revision of the four pillars is provided. However, the main contribution of the guide is the presentation of a strategy to include gender in the contents of an engineering subject. It is based on the classification of all the engineering subjects in three clusters: (1) scientific foundations, (2) technologies, and (3) management subjects. Within the scientific foundation cluster, the contextualization of problems and exercises (also for the assessment) is crucial. New contexts could be related to health and environmental. Even the data gathered in matrices in Algebra or Statistics could be gender segregated data and provide an interesting framework for a debate. An interesting example in this direction is written by Alsina, on the integration of gender dimension in STEM degrees (Alsina, 2019). The 17 Sustainable Development Goals, adopted as part of the 2030 Agenda for Sustainable Development, conform an interesting framework for the technologies cluster of subjects. Indeed, aspects such as perception of risk, ergonomics, environmental impact, consumption patterns, and others can be included as design parameters and, therefore, present gender differences can be analyzed and discussed. Also, service learning activities are a good strategy to increase female motivation and deal with communication skills and teamwork. An example of it can be found in reference (Calvo-Iglesias, 2016), where female engineering students prepared workshops to be hold in schools. In this direction, some research show that these role model interventions addressed at girls from and above 12-year-old are a way of reducing stereotypes and increasing girls' engagement with STEM fields (González-Pérez et al., 2020). Finally, in the management subjects' cluster the aspects that are gender-related are interpersonal skills, leadership, recruitment, and labor market. Concerning leadership, it relies on a more opened and cooperative point of view not only for women but also allowing a space for men to develop their own styles and a diversity of voices (López-Belloso et al., 2021).

9.4   Discussion

As we have seen in this article, STEM disciplines have similarities and differences. We have a problem of female vocations that is common to all of them and that we should address from an early age. However, at university level we can definitely contribute to promote equality among our students through the introduction of gender perspective in our STEM disciplines. By doing so, we can contribute to the reduction of the stereotypes, we can increase our students' awareness on the gender issues already occurring in STEM professions and we can enable them to become gender bias free citizens.

Being highly masculinized disciplines, we need to mainstream the gender perspective and comply with European and state regulations that demand it, but we also need to introduce specific subjects that train students in gender, at least while the level of mainstreaming is low. It is important to point out that we currently have guides for the introduction of the gender perspective that cover many of the STEM disciplines and, thanks to the equality units, numerous courses have been given to train university professors. However, to attract faculty to these courses we need incentive. We should not forget that in general, the academic career values research more, so it should be included as a merit in the accreditations to professors, for example. For all this we need to introduce changes in the evaluation agencies that promote changes in universities and allow us to overcome possible resistance to the introduction of the gender perspective in STEM degrees.

Other alternatives may be some of those discussed above, for example, being able to provide real-life application examples, or an education oriented to the fulfillment of the SDGs in engineering. We also consider very important activities such as the ones carried out in the framework of the European project *Diversity in the culture of physics* (Erlemann et al., 2019) or the creation of a virtual itinerary of training in gender, with courses like those of the UJI (Epifanio et al., 2021). The online activities also facilitate the reconciliation of work and personal life and could be a first step toward obtaining a certificate in equality.

9.5   Conclusions

In this communication we have shown the similarities and differences in relation to gender in Physics, Engineering, and Mathematics, as well as some strategies to integrate the gender perspective in these disciplines: a summer school on physics and gender, an asynchronous virtual course on mathematical coeducation or pilot projects carried out at the Universitat Politècnica de Catalunya (UPC) and Universitat de Barcelona (UB). Also, the guides for the incorporation of gender in university teaching, which show us how to integrate this perspective in the objectives and contents of the courses, the references and the teaching and evaluation methods.

Although awareness initiatives and a strong commitment from individual professors can be a good starting point, we cannot underestimate the power structures opposing real change in university settings. In fact, it is when trying to challenge these power structures that we realize how deeply entrenched they are.

A decided political action from institutions is needed, as we see for instance in the Catalan context, where the requirement of the assessing agency of University quality (AQU) of gender mainstreaming in all studies has led to a new concern for gender issues among faculties. How superficial this concern is going to be in the near future remains an open question.

In this sense, it is imperative to network, collaborate, and engage in debate among the different agents committed to making and teaching better science and technology for everyone.

References

Ahmed, S. (2017). *Living a feminist life*. Duke university Press.

Alsina, M., et al. (2019). STEM Students' Perception Of Gender Mainstreaming In Teaching: The development of a measuring tool. In *Proceedings of the International Conference of Education, Research and Innovation, Seville, Spain, 11–13 November 2019*.

Alsina, M. (2019). On the integration of gender dimension in STEM degrees: reflections and analysis, statistical activities. In *Proceedings of the International Conference of Education, Research and Innovation, Seville, Spain, 11–13 November 2019*.

Archer, L., et al. (2020). Learning that physics is 'not for me': Pedagogic work and the cultivation of habitus among advanced level physics students. *Journal of the Learning Sciences*, 1–38.

Ariño, A., et al. (2019). Via Universitària: Accés, condicions d'aprenentatge, expectatives i retorns dels estudis universitaris (2017–2019). Xarxa Vives: Valencia.

Barad, K. (2007). *Meeting the universe halfway*. Duke university Press.

Barrios, B., et al. (2021). STOP Discriminación: evaluación de la aplicación de la Ley 14/2011 de la ciencia en lo referente a la perspectiva de género con propuestas de mejora. In *Proceedings of the VII Xornada Universitaria Galega en Xénero, Galicia, 10–11 June 2021* (in press).

Birenbaum, M., & Feldman, R. A. (1998). Relationships between learning patterns and attitudes towards two assessment formats. *Educational Research, 40*(1), 90–98. https://doi.org/10.1080/0013188980400109.

Botella, C., Rueda, S., López-Iñesta, E., & Marzal, P. (2019). Gender diversity in STEM disciplines: A multiple factor problem. *Entropy, 21*(30), 1–17. https://doi.org/10.3390/e21010030.

Brage, T. (2019). What does gender have to do with physics? Optics & Photonics News. Retrieved July 18, 2021, from https://www.osa-opn.org/home/articles/volume_30/february_2019/departments/what_does_gender_have_to_do_with_physics/.

Brewe, E., et al. (2010). Toward equity through participation in modeling instruction in introductory university physics. *Physical Review Special Topics—Physics Education Research, 6*(010106), 1–12.

Buell, C. A., & Shulman, B. (2019). An Introduction to Mathematics for Social Justice. PRIMUS29: 205–209.

Cabero, I., & Epifanio, I. (2021). A data science analysis of academic staff workload profiles in Spanish universities: gender gap laid bare. *Education Sciences*. https://doi.org/10.3390/educsci11070317.

Calvo-Iglesias, E. (2020). Preparing biographies of STEM women in the wikipedia format, a teaching experience. *IEEE Revista Iberoamericana De Tecnologias Del Aprendizaje, 15*(3), 211–214. https://doi.org/10.1109/RITA.2020.3008144.

Calvo-Iglesias, E. (2021). Guide of Physics to mainstreaming gender in university teaching. Xarxa Vives d'Universitats, Castelló de la Plana. Retrieved July 18, 2021, from https://www.vives.org/book/guide-of-physics-to-mainstreaming-gender-in-university-teaching/.

Calvo-Iglesias, E. (2016). Una estudiante de ingeniería en cada cole. In Proceedings of VII Congreso Nacional y II Internacional de Aprendizaje-Servicio Universitario, Santiago de Compostela, 13–15 October 2016.

Carlone, H. B. (2004). The cultural production of science in reform-based physics: Girls' access, participation, and resistance. *Journal of Research in Science Teaching, 41*, 392–414.

Catalan University Quality Assurance Agency. (2018). General framework for the incorporation of the gender perspective in university teaching. Retrieved September 10, 2021, from https://www.aqu.cat/en/universities/Guies-metodologuiques/general-framework-for-incorporating-the-gender-perspective-in-higher-education-teaching.

Danielsson, A. T. (2012). Exploring woman university physics students "doing gender" and "doing physics." *Gender and Education, 24*(1), 25–39.

Epifanio, I. (2020). *Guía para una docencia universitaria con perspectiva de género de Matemáticas*. Xarxa Vives d'Universitats.

Epifanio, I., Ferrando, L., & Martínez-García, M. (2021). Mainstreaming gender in mathematics university teaching and an assessment from students and teachers. In *Proceedings of the XI Jornadas Internacionales de CAMPUS VIRTUALES, Salamanca, 30 Set 2021*.

Erlemann, M., et al. (2019). *Diversity in the cultures of physics: A European summer school curriculum*. Freie Universität Berlin.

Estradé, S. (2021). *Guia per a una docencia universitària amb perspectiva de gènere: Enginyeria Electrònica*. Xarxa Vives d'Universitats. In press.

Figures, S. H. E. (2019). *European commission, directorate-general for research and innovation*.

García-Dauder, S., & Pérez-Sedeño, E. (2017). *Las 'mentiras' científicas sobre las mujeres*. Catarata.

Gonsalves, A. (2014). "Physics and the girly girl—there is a contradiction somewhere": Doctoral students' positioning around discourses of gender and competence in physics. *Cultural Studies in Science Education, 9*, 503–521.

González-Pérez, S., Mateos de Cabo, R., & Sáinz, M. (2020). Girls in STEM: Is it a female role-model thing? *Frontiers in Psychology, 11*, 1–21. https://doi.org/10.3389/fpsyg.2020.02204.

Gónzalez-González, C., García-Holgado, A., & García-Peñalvo, F. J. (2019). Introduciendo la perspectiva de género en la enseñanza universitaria: co-creación de guías docentes y proyectos de innovación. In *Proceedings of IX Jornadas Internacionales de Campus Virtuales, Popayán, Colombia, 11–13 Setember 2019*.

Hofer, S. (2015). Studying gender bias in physics grading: The role of teaching experience and country. *International Journal of Science Education, 37*(17), 2879–2905. https://doi.org/10.1080/09500693.2015.1114190.

Hughes, G. (2001). Exploring the availability of student scientist identities within curriculum discourse: An anti-essentialist approach to gender-inclusive science. *Gender and Education, 13*(3), 275–290.

Johansson, A., & Berge, M. (2020). Lecture jokes: Mocking and reproducing celebrated subject positions in physics. In *Physics Education and Gender*. Springer Nature Switzerland.

Keller, E. F. (1985). *Reflections on gender and science*. Yale University Press.

Linkova, M., & Mergaert, L. (2021). Negotiating change for gender equality: Identifying leverages, overcoming barriers. *Investigaciones Feministas 12*(2), 297–308. https://doi.org/10.5209/infe.72319.

López-Belloso, M., Silvestre, M., & García-Muñoz, I. (2021). Igualdad de Género en instituciones de educación superior e investigación. *Investigaciones Feministas, 12*(2), 263–270. https://doi.org/10.5209/infe.76643.

Mas de les Valls, E., & Peña, M. (2020). Enginyeria Industrial. Guies per a una Docència Universitària amb Perspectiva de Gènere. Xarxa Vives d'Universitats, Castelló de la Plana.

Moss-Racusin, C. A., Dovidio, J. F., Brescoll, V. L., et al. (2012). Science faculty's subtle gender biases favor male students. *Proceedings of the National Academy of Sciences of the United States of America, 109,* 16474–16479.

Pekkarinen, T. (2015). Gender differences in behaviour under competitive pressure: Evidence on omission patterns in university entrance examinations. *Journal of Economic Behavior & Organization, 115,* 94–110. https://doi.org/10.1016/j.jebo.2014.08.007.

Peña, M., Olmedo-Torre, N., Mas de les Valls, E., & Lusa, A. (2021). Introducing and evaluating the effective inclusion of gender dimension in STEM higher education. *Sustainability, 13,* 1–26. https://doi.org/10.3390/su13094994.

Rasooli, A., Zandi, H., & DeLuca, C. (2018). Re-conceptualizing classroom assessment fairness: A systematic meta-ethnography of assessment literature and beyond. *Studies in Educational Evaluation, 56,* 164–181. https://doi.org/10.1016/j.stueduc.2017.12.008.

Riener, G., & Wagner, V. (2017). Shying away from demanding tasks? Experimental evidence on gender differences in answering multiple-choice questions. *Economics of Education Review, 59,* 43–62. https://doi.org/10.1016/j.econedurev.2017.06.005.

Schiebinger, L. (2021). Gendered innovations: integrating sex, gender, and intersectional analysis into science, health & medicine, engineering, and environment. *Tapuya: Latin American Science, Technology and Society 4*(1), 1–17. https://doi.org/10.1080/25729861.2020.1867420.

Sistema Integrado de Información Universitaria (SIIU). (2019). Dirección General de Universidades.

StoneWall. (2017). Creating an LGTB-inclusive curriculum. A guide for secondary schools. https://www.stonewall.org.uk/system/files/inclusive_curriculum_guide.pdf. Last accessed 18 July 2021.

Sáinz, M., et al. (2020). Gendered motivations to pursue male-dominated STEM careers among Spanish young people: A qualitative study. *Journal of Career Development, 47*(4), 408–423. https://doi.org/10.1177/0894845318801101.

Sáinz, M. (2011). Factors which influence girl's orientations to ICT subjects in schools. *Evidence from Spain. International Journal of Gender, Science and Technology 3*(2), 387–406.

Sáinz, M., et al. (2017). "¿Por qué no hay más mujeres STEM? Se buscan ingenieras, físicas y tecnólogas". Fundación Telefónica. Editorial Ariel. Barcelona.

Tannenbaum, C., Ellis, R. P., Eyssel, F., et al. (2019). Sex and gender analysis improves science and engineering. *Nature, 575,* 137–146.

Unidad de Mujeres y Ciencia. (2021). Científicas en cifras 2021. Ministerio de Ciencia e Innovación. Retrieved September 10, 2021, from https://www.ciencia.gob.es/site-web/Secc-Servicios/Igualdad/cientificas-en-cifras.html.

Van den Brink, M., & Stobbe, L. (2014). The support paradox: Overcoming dilemmas in gender equality programs. *Scandinavian Journal of Management, 30*(2), 163–174.

Verge, T. (2021). Gender equality policy and universities: Feminist strategic alliances to re-gender the curriculum. *Journal of Women, Politics & Policy, 42*(3), 191–206. https://doi.org/10.1080/1554477X.2021.1904763.

Verge, T., & Cabruja, T. (2017). *La perspectiva de gènere en docència i recerca a les universitats de la Xarxa Vives: Situació i reptes de futur.* Xarxa Vives d'Universitats.

World Economic Forum. (2016). The industry gender gap. Women and work in the fourth industrial revolution. Retrieved September 10, 2021, from http://www3.weforum.org/docs/WEF_FOJ_Executive_Summary_GenderGap.pdf.

Chapter 10
Examples of Good Practices in Erasmus+ Projects that Integrate Gender and STEM in Higher Education

M. Goretti Alonso de CastroⒹ **and Francisco José García-Peñalvo**Ⓓ

Abstract The European Union promotes the realization of European projects through different programs, among which the Erasmus+ program stands out in the educational field. These projects deal with different topics, including gender and STEM in the different educational sectors, Higher Education included. In addition, the projects are cataloged and can be searched and consulted in the Erasmus+ Project Results Platform that allows filtering and choosing by educational sector, project type, by specific tokens, as well as selecting those that are labeled as success stories and/or good practices. Knowing which have been the projects that have been considered the most outstanding, successful, and/or good practice in the field of gender and STEM in Higher Education and finding the aspects that characterize them can be a source of inspiration to deepen in this field of research based on current experiences that have worked. This is the purpose of this chapter, which shows these outstanding projects as references in the field of STEM empowerment among women in Higher Education.

Keywords STEM · Gender · European Projects · Erasmus+ · Education

10.1 Introduction

The society in which we live is diverse and as such it is necessary to meet the educational needs of all groups and ensure that everyone feels included. It is necessary to train all citizens to be competent for the development of an adequate professional and personal life. This entails working on equity and diversity in our educational centers and is one of the principles of the educational law in Spain (BOE, 2020). In addition, this same law also considers gender equality as one of the objectives in all

M. G. A. de Castro (✉)
Education in the Knowledge Society PhD (GRIAL Research Group), University of Salamanca, Salamanca, Spain

F. J. García-Peñalvo
GRIAL Research Group, Computer Science Department, Research Institute for Educational Sciences, University of Salamanca, Salamanca, Spain
e-mail: fgarcia@usal.es

© The Author(s) 2022
F. J. García-Peñalvo et al. (eds.), *Women in STEM in Higher Education*, Lecture Notes in Educational Technology, https://doi.org/10.1007/978-981-19-1552-9_10

educational stages. UNESCO has defined a series of Sustainable Development Goals (SDGs) (UNESCO, 2021a) the fourth being for education. Within this objective, two of the goals seek equity, access, and democratization of higher education; the third and five targets are linked with gender equality, guaranteeing an inclusive and equitable quality education, and promoting lifelong learning opportunities for all. Specifically, gender equality is a cross-cutting objective present in most of the SDGs. The data compiled by UNESCO, through its Institute of Statistics (UNESCO, 2021b) shows a gender gap in professional careers related to Science, Technology, Engineering and Mathematics (STEM), both in an educational and professional context, therefore achieving work to reduce these differences is a great social challenge. These SDGs targets (UNESCO, 2021c) are as follows:

"Target 4.3: By 2030, ensure equal access for all women and men to affordable quality technical, vocational, and tertiary education, including university" (p. 1).

"Target 4.5: By 2030, eliminate gender disparities in education and ensure equal access to all levels of education and vocational training for the vulnerable, including persons with disabilities, indigenous people, and children in vulnerable situations" (p. 1).

An example of this need and challenge, regarding STEM and gender, is shown in compass brief number 13 published by the IEA in April 2021 on female science and mathematics teachers (Hastedt et al., 2021). This publication points out the need to make STEM teachers aware of their strengths and develop their self-efficacy based on the following findings:

"There is no direct relationship between the gender of the teacher and students' performance in science and mathematics. Grade 4 and 8 students taught by female teachers perform just as well in science and mathematics than their peers taught by male teachers. Yet, results show that female science and mathematics teachers have less self-efficacy than their male counterparts" (p. 1).

The European Union also works to achieve the SDGs and has among its goals the improvement of education, achieving greater equity, and attention to diversity. To this end, among other things, it promotes the implementation of European educational projects with funding aimed at improving teaching–learning systems (UE, 2021a), the Erasmus+ Programme (UE, 2021b) stands out in this area. This program finances educational projects at all levels, including higher education, so that institutions implement and explore new educational methodologies with projects that have among their priorities: inclusion and diversity, digital transformation, the environment and the fight against climate change and participation in democratic life.

In the Erasmus+ programme there is a database, known as Erasmus+ Project Results Platform (E+PRP) (UE, 2021c), in which a compilation of all the funded projects is available, identifying those that have been classified as good practice or success story. Therefore, reviewing successful projects in any field, and specifically in STEM and gender, can be of great help to see what practices are helping to reduce the gap in this educational and professional field.

Knowledge of successful projects can serve as a beacon to achieve, on the one hand, what is working well in other institutions and, on the other, detect possible needs to expand the scope of research.

For the revision of the projects in E+PRP, the methodology of systematic reviews of research projects (García Holgado et al., 2019b, 2020c) is really useful because it provides a perfect approach to analyze projects because it gives an overview of current trends, allowing the identification of gaps and opportunities. This methodology is the one that is being used in the research on "Methodological guide for the successful use of digital technologies in education: Improving learning through European educational projects" (Alonso de Castro & García-Peñalvo, 2020a, 2020b, 2021) in order to collect information on successful projects within the framework of Erasmus+ related to eLearning and the same technique is applied for the sample of projects that are presented in this chapter.

Within the framework of the GRIAL research group, in which the development of this chapter is carried out, some contributions that seek to understand the gender gap in higher STEM studies both in Spain and in Europe stand out (García-Holgado et al., 2019a, 2020a, 2020b; Verdugo-Castro et al., 2019, 2020a, 2020b, 2020c). There are several analyses that include intervention proposals, interviews, and case studies that give an idea of the importance of this topic for the group. Therefore, it is considered a topic of great relevance in the group, and this justifies analyzing practical examples of projects that have been considered good practice. For this reason, this chapter provides specific cases of projects classified as good practice in Erasmus+ Projects Platform and that have worked on STEM and gender in higher education so that we can take note of the factors that have been useful in the institutions involved. At the same time, the situation of projects of this type within the framework of Erasmus+ and the possibilities of future work will be analyzed.

In the analysis of projects that meet the established criteria (Erasmus+ , Good Practice, higher education, STEM, and gender), at this moment, a total of 5 projects have been found out of the total of 19 existing linked to STEM and gender and the more than 35.500 of higher education projects in Erasmus+ . In addition, there is another project labeled as good practice in higher education, which also works to improve the attraction to STEM careers, which although it is not focused on women, can be considered useful for them as well. Therefore, below we are going to see these six projects with the information gathered from E+PRP and the projects websites, followed with the main conclusions.

10.2 Project 1—Augmented Reality for Science Education

This project has been coordinated "Via University College" and had five partners: "University of Manchester", "Skolen I Midten", "Fundación Pública Gallega centro Tecnológico de Supercomputación de Galicia", "Hogskolen i Oslo og Akershus", and "Centro Público Integrado O Cruce". A total of six partners from higher education, IT and education, and the school sector from different countries (Denmark, Norway, United Kingdom, and Spain).

It has to do with improving the attractiveness of STEM subjects in schools. It is a transversal project that covers different educational levels, not only Higher Education, since as can be seen in the composition of partners there are both Universities and Foundations and educational centers. Therefore, it is a multi-stage project, in which Universities play an important role in researching results, but which is also applied in non-tertiary level schools.

In addition, it does not focus on the stimulation of women in the STEM field but tries to improve the attractiveness toward these areas in all sexes. As such, it is also valid for women, who through the use of technology, in this case augmented reality (AR), will be encouraged to see science from a practical point of view and will be encouraged to choose careers and professions related to the field of study. In short, it tries to raise awareness and promote a taste for STEM from the base of education to instill it from an early age. All the information about this project can be found on: https://ec.europa.eu/programmes/erasmus-plus/projects/eplus-project-det ails/#project/2014-1-DK01-KA200-000773 as well as on the project webpage http://www.ar-sci.dk/.

10.2.1 Project Context

This project focuses on meeting the need to work on new educational approaches and methodologies through new technologies as a means of improving teaching–learning processes. Specifically, it focuses on issues related to the teaching and learning of science and, in addition, to the teaching of the necessary skills in the twenty-first century in educational centers.

It is based on the fact that science subjects pose problems or obstacles for a large number of students in European schools. Science subjects are often considered "difficult" and require high levels of abstraction. This has led to a decrease in the interest of young Europeans in scientific subjects, both during their formative stage and as professional opportunities. The main goal of this project is that science education can be reinforced through the use of AR, through active and collaborative learning, as well as the interaction and visualization of central scientific knowledge. Furthermore, the technology is believed to have matured enough to introduce it into school contexts and involve teachers in the design and production of AR materials.

10.2.2 Project Objectives

The main goals of the project are to improve the quality of science teaching and learning processes with innovative methodologies; increase the motivation and attitude of students toward science education; achieve a student-centered model for science education, facilitating inquiry-based teaching, collaboration, and active learning; and strengthen and improve teaching and learning through technology, increasing the attractiveness and didactics for students and teachers.

It seeks not only to increase student participation but also to get teachers sufficiently trained in new technologies so that they get more involved with their students and this has been done using augmented reality technologies.

10.2.3   Project Results

The methodology developed in the project presents a student-centered approach. From the tests carried out in the educational centers, a high motivation on the part of the students in the use of technologies such as AR has been detected and this helped them to understand the most abstract concepts. In addition, it was possible to attract more attention to STEM topics. Specifically, 76% of the students stated that they learned through collaboration with their peers and 60% observed a different role in their teaching staff. After the third round of testing in educational centers, 75% of the students, who participated in the tests, indicated that their interest in STEM subjects had increased.

The project has made it possible to foster collaboration between schools, teachers, students, teacher trainers, researchers, and developers of educational technology.

The main results for this project have been: a user guide, webinars on AR in science teaching, needs analysis scientific articles, materials development, guide for the uses of AR in science education, 6 Piloting and pilot reports, and a full list of AR-materials.

10.3   Project 2—Engendering STEM

The project is a partnership coordinated by "City of Glasgow College" with five partners: "Instituto Específico de Formación Profesional Superior Miguel Altuna", "Stichting VHTO" and "Edinburgh Napier University" from three countries: United Kingdom, Spain, and Netherlands. All of them have a link with the promotion of gender equality within the STEM field and belong both to the field of education at different educational levels and also in the workplace.

The focus of the project is on the growing gap in professional training and employment in the Science, Technology, Engineering, and Mathematics fields, within the framework of the European Union. In addition, it highlights that female participation in the labor market is very low in these areas. Therefore, it seeks to develop a qualified workforce in STEM to meet the demands of the labor market by increasing and involving the female population. This implies overcoming the barriers perceived by this population. One of the keys is to empower employers and teachers to improve their work practices in order to achieve more inclusion and attractiveness on that field. The information of the project is compiled on: https://ec.europa.eu/programmes/erasmus-plus/projects/eplus-project-

details/#project/2017-1-UK01-KA203-036834, and the project website https://www.
engenderingstem.co.uk/).

10.3.1 Project Context

The main goal of the project is to recognize the factors that distinguish small and
medium-sized companies (SMEs) from STEM that have achieved equality and diversity in the workplace and those that still have large differences. Specifically, it sought
to identify effective strategies to improve gender equality in institutions, as well as
the different forms and stages of participation.

As an added target of the project, it highlights that it sought to achieve a positive impact on the hiring of women in the STEM sector, especially in the area of
SMEs. Additionally, they proposed to support the personnel selection and talent
retention processes in this area. In turn, provide educational resources to improve
understanding and knowledge of relevant issues in the field. All with a free or low-cost
flexible educational content approach.

To achieve this, the project has worked on the development of a self-assessment
toolkit, together with some good practice guides in increasing gender equality, as
well as a tailor-made training program.

10.3.2 Project Objectives

The most outstanding objectives of the project have been:

- Achieve better knowledge and awareness of gender equality within the STEM
  sector by training as many participants as possible.
- Involve agents and institutions related to the field of work in online activities
  throughout the project, fostering interactions on social networks, participation to
  create best practice guides and other project results.
- Seek support from SMEs to test and assess the validity of the Toolkit for the self-
  assessment of gender equality, reviewing the hiring, retention, and promotion
  policies and practices, as well as in the writing of the personalized evaluation
  report according to their specific needs.
- An additional objective was to increase the average number of women hired by
  participating SMEs.

10.3.3 Project Results

The project delivered has developed three key products:

- Publication of the research carried out on gender equality, as well as the guides of the best practices detected.
- Online Diagnostic Toolkit for Self-Assessment of Gender Equality.
- Blended learning training program.

Through these products generated in the project, the following milestones have been achieved:

- 1,191 participants have been trained in the STEM sector on actions to be carried out for gender equality in this field.
- Involve 2,587 institutions for the development of resources, guides, and interaction in social networks and website.
- Support 56 employers of STEM SMEs in the use of the self-assessment Toolkit for practices and policies applied in relation to gender equality in their companies.
- 47% of the employers who participated in the project have increased the number of women hired in their organizations.

In addition, the project guarantees long-term sustainability through the development of networks of employers, trainers, teachers, and researchers, as well as the implementation of all the tools generated within the procedures used by the project partners.

10.4 Project 3—Innovative Women Entrepreneurs of the Future

This project is a strategic partnership in the field of Higher Education in which the coordinator is "Bursa Technical University" that worked with five partners, "Politecnico di Torino", "I3P—The Innovative Companies Incubator of Politécnico di Torino", "Silesian University of Technology", "Technopark Gliwice", and "KOSGEB—Small and Medium Enterprises Development Organization". The organizations are from three different countries: Turkey, Italy, and Poland have varied profiles of the educational and business field connected to STEM. The information of the project is available on https://ec.europa.eu/programmes/erasmus-plus/projects/eplus-project-details/#project/2016-1-TR01-KA203-035231 and also on the project website https://www.innowoment.org/.

Additionally, this project is included on the research work "Methodological guide for the successful use of digital technologies in education: Improving learning through European educational projects" (Alonso de Castro & García-Peñalvo, 2020a).

10.4.1 Project Context

The project is based on statistical data on women entrepreneurs in Europe which indicates that women only represent one third of all entrepreneurs in science and technology sectors. This in turn has to do with the proportion of female students studying science, technology, engineering, and mathematics in higher education. Precisely for this reason, among the priority areas for higher education institutions of the EU Modernization Agenda is the stimulation of entrepreneurial capacity through learning in interactive environments and improving their ability to participate in the creation of companies.

10.4.2 Project Objectives

The following objectives of the project stand out:

- Raise awareness about the gender gap in technological fields and encourage more women to study Science, Technology, Engineering, and Mathematics.
- Train in transversal and entrepreneurial competences, such as critical thinking, problem solving, creativity, analysis, knowledge of digital tools, and foreign languages. For which, courses, seminars, workshops, and mobilities for learning will be developed within the framework of the Erasmus+ programme.
- Develop training materials for all potential entrepreneurs.
- Establish a roadmap to achieve an efficient business ecosystem so that potential entrepreneurs can participate in the training network, mentoring and with the support they need.

10.4.3 Project Results

The results of the project have been compiled into two books.

- A book on 'Experiences of innovative women entrepreneurs of the future', which is a complete guide not only for women, but also for all people who want to start their own companies.
- An electronic book entitled 'A roadmap for a successful incubator' has also been implemented, which seeks to serve as a manual not only for future entrepreneurs, but also all institutions that wish to support business creation.

Apart from those books, video clips have been developed in order to encourage more female students to study Science, Technology, Engineering, and Mathematics degrees, as well as to promote the creation of their own companies. Additionally, it was provided an Online Learning Toolkit.

10.5 Project 4-Euro4Science: Exploring "CSI Effect" and Forensic Sciences to Boost the Appeal of Science to Young People and Reinforce Interdisciplinarity in European High Schools

The project is a partnership coordinated by the University of Aveiro (Portugal) and in which two schools (in Portugal and the United Kingdom), an NGO (Bulgar-ia) and 2 SMEs (Portugal and Poland) participated. Institutions with varied profiles that add value to the purposes of the project.

The project focuses on the problems derived from early school leaving in addition to the need to improve the attractiveness of science, technology, engineering, and mathematics for young people, which is key to having people trained for the needs of the world of work. To do this, it takes advantage of the fact that young people are particularly open to the Crime Scene Investigation (CSI) iconography to use this theme as a strategy to reduce early school drop-out, as well as to promote cultural exchanges and interdisciplinarity to improve social inclusion. The information of the project is available on: https://ec.europa.eu/programmes/erasmus-plus/projects/eplus-project-details/#project/2014-1-PT01-KA200-001012 and the project website http://euro4science2.eu/.

10.5.1 Project Context

Euro4Science's main objective was to implement innovative educational practices, tools and methodologies to improve the quality of the teaching–learning process with regard to the attractiveness of science-related subjects and careers.

10.5.2 Project Objectives

Among the most relevant objectives of the project are as follows:

- Encourage interdisciplinary pedagogical approaches, with the collaboration in the implementation of an Educational Toolkit in Forensic Sciences.
- Participation of teachers in experimental education by testing and using the Toolkit and participating in various activities of the project.
- Increase the motivation of European students toward culture and scientific professions through participation in scientific activities with experts.
- Contribute to reducing early school leaving due to greater motivation for teaching and a broader knowledge of possible career options, including those related to scientific fields.

- Pay particular attention to the messages conveyed to girls, to increase attractiveness toward science subjects and degrees in science-related fields.

10.5.3   Project Results

Euro4Science, managed to increase the attractiveness of scientific subjects through motivational activities linked to the "CSI Theme", as well as involving teachers and students of different ages at an international level. This was achieved through:

- A forensic science education toolbox in 4 languages (English, Portuguese, Bulgarian, and Polish) was implemented and tested.
- 3 "CSI weeks" with motivating activities for students and with the participation of the entire educational community.
- Workshops for teachers; exchanges between groups of students and international conferences "CSI @ school".

10.6   Project 5—Early Identification of STEM Readiness and Targeted Academic Interventions

This project is a partnership led by the University of Leuven (KU Leuven) with three key partners (Hamburg University of Technology [Germany], University of Žilina [Slovakia] and KU Leuven [Belgium]), in addition to three supporting partners (Budapest University of Technology and Economics [Hungary], Aalto University [Finland] and University of Birmingham [UK]) and a partner of the European network (European Society for Engineering Education—SEFI). The three key partners are STEM education research experts and the supporting partners provided relevant information and carried out case studies within their universities. In turn, having the SEFI network has made it possible to effectively disseminate the results throughout the project. As in the rest of the projects this variety of institutions enrich the work and provides a better impact in all the stakeholders.

The E+PRP URL for the project is https://ec.europa.eu/programmes/erasmus-plus/projects/eplus-project-details/#project/2014-1-BE02-KA200-000462 and more detail information could be found on the project website https://iiw.kuleuven.be/english/research/readystemgo.

10.6.1   Project Context

The project seeks to promote scientific and technological development in Europe through adequate training in secondary and higher education. Although it is true that there has been an advance in the enrollment of science, technology, engineering, and mathematics studies in most European countries, dropout rates are high, and it is necessary to work on the retention of students in these programs. Hence, the main goal of this project is to increase student retention rates in higher education STEM studies by early identification of those students at risk of dropping out. In addition, gender equity and equal opportunities are among its priorities.

10.6.2 Project Objectives

To achieve the main goal of the project, three objectives were established:

- Identification of the key competencies necessary to successfully pass the first year in a STEM program.
- Development of a comprehensive inventory with diagnostic tests that allow measuring different competencies, with an assessment of their predictive validity.
- Research on which intervention tools are best suited to help at-risk students and assess their efficiency.

10.6.3 Project Results

The first result was the holding of focus group discussions of first-year students, and some results were, for example, that first-year students found it difficult to adapt to the demands of higher education, as well as problems with understanding when they should apply reasoning based on formulas. Afterward, a quantitative survey was carried out on the experience of the first year and the obstacles encountered with 1,451 students from all the partner universities of the project.

A Learning and Study Strategies Inventory (LASSI), which also contained questions on educational background, was then provided to more than 9,000 first-year students and the results were linked to performance at the end of the first year. Students with poor time management skills were at higher risk of low performance or dropping out at the end of the first year. Educational background also had an impact on performance.

In addition, a Science Learning Attitudes Survey was conducted measuring the epistemological beliefs of more than 700 first-year students. The results were that 48% of the students indicated that they applied strategies based on formulas when solving engineering problems. Additionally, it was created an inventory with more than 100 diagnostic tests and concept inventories to measure academic and non-academic skills, all with a focus on STEM.

10.7 Project 6-Robotics Opportunities (to Foster) STEM Education

The project is based on an association led by "Universita degli Studi di Firenze" with the collaboration of eleven partners of multiple profiles and countries: "Ankara Ozel Tevfik Fikret Anadolu Lisesi", "Iis Ferraris Brunelleschi Empoli", "Instituto Politecnico do Porto", "Middle East Technical University", "Tartu Kivilinna Kool", "Istituto Comprensivo Don Lorenzo Milani", "Technische Universitaet Graz", "NPO Robootika", "Istituto Tecnico Tecnologico Statale Silvano Fedi Enrico Fermi.

Tartu Ulikool"and "Ufficio Scolastico Regionale per la Toscana".

The ROSE association is based on improving prospects given the current crisis context and youth unemployment in Europe, which contrasts with the difficulty of companies to hire engineers, technologists, and scientists. Therefore, the project addresses the improvement of STEM education using robotics and addresses the needs to reduce the gender gap by involving all genders, with special attention to women, and making science more attractive for all. The E+PRP URL is https://ec.europa.eu/programmes/erasmus-plus/projects/eplus-project-details/-project/2014-1-IT02-KA200-003,660 and the project webpage http://www.roseproject.eu/.

Furthermore, it is one of the projects included on the research work "Methodological guide for the successful use of digital technologies in education: Improving learning through European educational projects" (Alonso de Castro & García-Peñalvo, 2020a).

10.7.1 Project Context

The project focuses on improving the level of key competences, paying special attention to their relevance to the labor market, as well as promoting quality improvement, excellence in innovation and internationalization at an educational level, through increased transnational cooperation. All this on the basic objective that is based on enhancing the attractiveness of STEM disciplines using robotics and automation. It worked to make STEM disciplines more attractive to students from an early stage by informing them of the professional possibilities that can be offered in this area. On the other hand, focusing on the field of robotics and automation, will encourage them to continue studying. Gender, minorities, and accessibility issues are also tackled, seeking to cover a social need, since participation in STEM disciplines is not sufficiently balanced.

10.7.2   Project Objectives

The main objectives of the project are:

- Work at an educational level to improve employability and competitiveness through better training of the population in science, technology, engineering, and mathematics.
- Empower and improve the attractiveness of STEM subjects and professions at the school level.
- Development of student competition models, with special attention to younger students to whom these initiatives are not suitable, looking for alternatives in those cases.

10.7.3   Project Results

The products resulting from the project have been:

- A questionnaire and survey to analyze the starting point. The answers to the questionnaire are available on E+PRP and the project website.
- A questionnaire at the end of the project to measure the results and impact on the attractiveness gained in STEM areas within students. Answers are also available on E+PRP and the project website.
- The design of a vertical Curriculum in Educational Robotics for Italian Schools developed in Italian.
- Didactic Units and assessment rubrics for the vertical Curriculum in Educational Robotics.
- The design of a Vertical Curriculum in Educational Robotics for Schools—Summary in English.
- A video for the dissemination of ROSE Project.
- Rubric for assessment of class activity.
- Analysis report.

10.8   Conclusions

This chapter has presented six projects classified as good practice in the Erasmus+ project results platform that are related to STEM and five of them that also work on gender-related aspects. The projects are associations of institutions with different characteristics: Universities, non-tertiary educational centers, NGOs, SMEs, Foundations, etc. This variety of institutions provides great value since it allows working on important aspects of STEM and gender such as improving the attractiveness of STEM subjects and degrees through innovative methodologies, tools, seminars and workshops, networks, etc. as well as increase the number of women working in STEM.

All the projects analyzed on this chapter use of cutting-edge technologies such as augmented reality, robotics, technology in the forensic field, and online resources.

The projects also work on aspects related to improving attractiveness of STEM, entrepreneurship, and the reduction of early school leaving, key aspects to achieve an educational system more prepared to have trained people to meet the needs of the world of work.

Some characteristics common to all these projects, and that deserves to be highlighted as a reference for other projects, are as follows:

- The projects analyzed seek to work on motivation toward STEM professions within the educational field, creating interesting materials and resources for students from an early stage as well as guides and support materials for teacher training in improving teaching–learning processes.
- All have developed tangible resources that are available on their websites to the entire educational community. The fact of providing the materials in an open way and disseminating them among the stakeholders increases the impact capacity of the projects.
- The resources developed are aimed at the students and teachers who are going to use them, a fact that favors their implementation within the study programs of educational institutions as common tools. This will help to go on using them beyond the Erasmus+ funding period.
- All the resources implemented have been tested with participants related to the area of research: students, teachers, educational centers, and experts within the scope of STEM. Involving people identified as having a need for such resources or who have problems with their studies, and carrying out tests of their development is very useful to measure the degree of validity of the model that is developed. In addition, it guarantees that their use and improvement are maintained over time.
- Technology and innovation are important factors in all projects, as well as novel methodologies that can attract the target audience of each research. Technology not only helps in motivation but also allows the teaching–learning processes to be adapted to the real needs of everyone.
- Another notable feature is the networking with educational institutions and other organizations that complement and help to implement new technologies or provide experts in the field of study. Networking encourages collaboration, exchange of good practices, and, at the same time, it favors dissemination among peers, increasing the impact and real implementation of the products that are developed in the projects within the institutions.
- Online learning, either with eLearning or in a blended mode, is a general trend in all the projects reviewed. In fact, two of the projects included in this sample of good practices are linked with eLearning and are part of the research "Methodological guide for the successful use of digital technologies in education: Improving learning through European educational projects" (Alonso de Castro &

García-Peñalvo, 2020a, 2020b, 2021), research in which good practice projects in Erasmus+ are also analyzed.

Additionally, in relation to the number of existing Erasmus+ projects in higher education on the subject discussed in this chapter, there is a lot of room for innovation and development of educational resources through the Erasmus+ Programme in this educational field within the areas of STEM and gender. Only 19 Higher Education projects have explicitly worked on STEM and gender and knowing that there are more than 35,500 Erasmus+ projects in Higher Education.

In summary, even knowing that there are only few examples of good practices available in higher education under the umbrella of Erasmus+ , the ones that exist are a good example that serves as inspiration to review those that have had a great impact and attend the needs of higher education institutions in the field of STEM and gender.

Acknowledgements This research work has been carried out within the University of Salamanca Ph.D. Programme on Education in the Knowledge Society scope (http://knowledgesociety.usal.es) (García-Peñalvo, 2013) with the mentoring and supervision of Francisco José García-Peñalvo, as well as the support available from the University of Salamanca and specifically from the GRIAL group (García Peñalvo et al., 2019; Grupo GRIAL, 2019).

References

Alonso de Castro, M. G., & García-Peñalvo, F. J. (2020a). Methodological guide for the successful use of digital technologies in education: Improvement of learning through European educational projects. In *Eighth International Conference on Technological Ecosystems for Enhancing Multiculturality (TEEM'20), October 21–23, 2020a, Salamanca, Spain.* New York, USA: ACM. https://doi.org/10.1145/3434780.3436549.

Alonso de Castro, M. G., & García-Peñalvo, F. J. (2020b). Overview of European educational projects on eLearning and related methodologies: Data from Erasmus+ project results platform. In *Eighth International Conference on Technological Ecosystems for Enhancing Multiculturality (TEEM'20), October 21–23, 2020b, Salamanca, Spain* (p. 8). New York, USA: ACM.https://doi.org/10.1145/3434780.3436550.

Alonso de Castro, M. G., & García-Peñalvo, F. J. (2021). Erasmus+ educational projects on elearning and related methodologies: Data from Erasmus+ project results platform. In F. J. García-Peñalvo (Ed.), *Information Technology Trends for a Global and Interdisciplinary Research Community* (pp. 111–133). IGI Global. https://doi.org/10.4018/978-1-7998-4156-2.ch006.

BOE. (2020). Ley Orgánica 2/2006, de 3 de mayo, de Educación. Retrieved September 25, 2021, from https://www.boe.es/buscar/act.php?id=BOE-A-2006-7899.

García-Peñalvo, F. J. (2013). Education in knowledge society: A new Ph.D. programme approach. In F. J. García-Peñalvo (Ed.), *Proceedings of the First International Conference on Technological Ecosystems for Enhancing Multiculturality (TEEM'13) (Salamanca, Spain, November 14–15, 2013). ACM International Conference Proceeding Series (ICPS)* (pp. 575–577). New York, USA: ACM.

García-Holgado, A., Verdugo-Castro, S., González, C., et al. (2020a). European proposals to work in the gender gap in STEM: A systematic analysis. *IEEE Revista Iberoamericana De Tecnologias Del Aprendizaje, 15,* 215–224.

García-Holgado, A., et al. (2020b). Facilitating access to the role models of women in STEM: W-STEM mobile app. In *7th International Conference, LCT 2020, Held as Part of the 22nd HCI International Conference, HCII 2020*.

García Holgado, A., Marcos-Pablos, S., & García-Peñalvo, F. J. (2020c). Guidelines for performing systematic research projects reviews. *International Journal of Interactive Multimedia and Artificial Intelligence, 6*(2), 136–144. https://doi.org/10.9781/ijimai.2020.05.005.

García-Holgado, A., et al. (2019a). Trends in studies developed in Europe focused on the gender gap in STEM. In *Proceedings of the XX International Conference on Human Computer Interaction*. New York, USA: ACM. https://doi.org/10.1145/3335595.3335607.

García Holgado, A., Marcos-Pablos, S., Therón-Sánchez, R., et al. (2019b). Technological ecosystems in the health sector: A mapping study of European research projects. *Journal of Medical Systems, 43*(art. 100), 2019. https://doi.org/10.1007/s10916-019-1241-5.

García Peñalvo, F. J., et al. (2019). Grupo GRIAL. IE Comunicaciones. *Revista Iberoamericana de Informática Educativa* (30), 3348.

Grupo GRIAL. (2019). Producción Científica del Grupo GRIAL de 2011 a 2019 (GRIALTR2019010). Grupo GRIAL, Universidad de Salamanca, Salamanca, España. Retrieved from https://repositorio.grial.eu/bitstream/grial/1624/1/GRIAL-TR-2019-010.pdf.

Hastedt, D., et al. (2021, April) *Female science and mathematics teachers: Better than they think?* (IEA Compass: Briefs in Education No. 13). Amsterdam, The Netherlands: IEA.

UE. (2021a). *Unión Europea. Erasmus+*. Retrieved September 25, 2021, from https://ec.europa.eu/programmes/erasmus-plus/node_es.

UE. (2021b). *Unión Europea. Guía del Programa Erasmus+*. Retrieved September 25, 2021, from https://ec.europa.eu/programmes/erasmus-plus/resources/programme-guide_es.

UE. (2021c). *Unión Europea. Plataforma de Resultados de Proyectos Erasmus+*. Retrieved September 25, 2021, from https://ec.europa.eu/programmes/erasmus-plus/projects/.

UNESCO. (2021a). *UNESCO and sustainable development goals*. Retrieved September 25, 2021, from https://en.unesco.org/sustainabledevelopmentgoals.

UNESCO. (2021b). *UNESCO institute for statistics database*. Retrieved September 25, 2021, from http://data.uis.unesco.org/.

UNESCO. (2021c). *Sustainable development goal 4 and its targets*. Retrieved September 25, 2021, from https://en.unesco.org/education2030-sdg4/targets.

Verdugo-Castro, S., et al. (2019). Revisión y estudio cualitativo sobre la brecha de género en el ámbito educativo STEM por la influencia de los estereotipos de género. In A. Pedro Costa, I. Pinho, B. M. Faria & L. P. Reis (Eds.), Atas - Investigação Qualitativa em Ciências Sociais/Investigación Cualitativa en Ciencias Sociales, vol. 3, pp. 381–386.

Verdugo-Castro, S., et al. (2020a). Análisis e intervención sobre la brecha de género en los ámbitos educativos STEM. In Estudios interdisciplinares de género, 1ª ed., pp. 591–608.

Verdugo-Castro, S., et al. (2020b). Interviews of Spanish women in STEM: a multimedia analysis about their experiences. In *8th International Conference on Technological Ecosystems for Enhancing Multiculturality (TEEM 2020b)*.

Verdugo-Castro, S., et al. (2020c). Pilot study on university students' opinion about STEM studies at higher education. In *Eight International Conference on Eight International Conference on Technological Ecosystems for Enhancing Multiculturality (TEEM 2020c) Ecosystems for Enhancing Multiculturality (TEEM 2020)*.